I0908077

Shortwave Listener's Guide

by

H. Charles Woodruff

HOWARD W. SAMS & CO., INC.
THE BOBBS-MERRILL CO., INC.
INDIANAPOLIS · KANSAS CITY · NEW YORK

FIFTH EDITION

FIRST PRINTING—1973

International Standard Book Number: 0-672-20934-9
Library of Congress Catalog Card Number: 72-88695

Preface

Every owner of a shortwave receiving set is familiar with the thrill that comes from hearing a distant station broadcasting from a foreign country. To hundreds of thousands of people the world over, shortwave listening (often referred to as swl) represents the most satisfying, the most worthwhile of all hobbies.

A recently conducted survey disclosed that more than 25 million shortwave receivers are in the hands of the American public, with the number increasing daily. To explore the international shortwave broadcasting bands in a knowledgeable manner, the shortwave listener must have available a list of shortwave stations, their frequencies, and their times of transmission. To keep abreast of the ever-increasing public interest in music, news, and the exchange of cultural ideas from foreign lands, the fifth edition of this book has again been completely revised to include the most recent changes in broadcasting schedules. The listings are conveniently arranged in four sections to help the swl'er more fully enjoy his hobby.

Section 1 consists of worldwide shortwave broadcasting stations listed alphabetically according to country and location within the country. The important particulars such as call letters (when assigned), rf carrier output in kilowatts (kW), frequency in megahertz (MHz), and hours of transmission (in Eastern Standard Time) for each station are given.

Section 2 contains a listing of shortwave broadcasting stations in numerical order by frequency, including the location and country.

Section 3 is divided into six parts, each titled with its respective time period, such as Midnight—4:00 am EST; 4:00 am—8:00 am EST; 8:00 am—Noon EST; Noon—4:00 pm EST; 4:00 pm—8:00 pm EST; 8:00 pm—Midnight EST. Within each of the six subdivisions, the shortwave broadcasting stations are listed in alphabetical order by country and location. The actual transmission time within the respective period is also given.

Section 4 contains a listing of clandestine radio broadcasting stations that have been heard periodically. These stations either operate behind the Iron Curtain or are communist-controlled stations operating in free countries. The times of transmission and station frequencies have been logged as they have been heard. For obvious reasons, no assurance can be given that the stations will be received at the time or on the frequency listed.

The tabulations in this book by no means represent all of the shortwave broadcasting stations in the world. Only those normally heard in the United States are included. The accuracy cannot be guaranteed; carrier frequencies and program scheduling may change without notice. However, every effort will be made to increase its usefulness by periodic updating. All swl'ers are cordially invited to comment on any additions, deletions, or changes that may be noted.

H. Charles Woodruff

Contents

Introduction

To pursue the very interesting and stimulating hobby of shortwave listening in an informed manner, the hobbyist must be aware of a few salient facts. These important items are discussed in the following paragraphs. Every effort has been made to simplify the data. If detailed information on a particular subject is desired, it is suggested that a textbook be consulted.

Frequency

All transmission frequencies listed in this book are expressed in megahertz (MHz). Although the frequencies listed are carried to three decimal places for consistency, the tuning dial of most shortwave receivers will omit the decimals. For example, the numerals 9, 10, 15, 20, etc., appearing on the tuning dial stand for 9 megahertz (9 MHz), 10 megahertz (10 MHz), 15 megahertz (15 MHz), and 20 megahertz (20 MHz). To determine the location of a station broadcasting on 9.100 megahertz, the operator need only to mentally divide the space between 9 MHz and 10 MHz, and position the receiver tuning-dial marker one-tenth of that spacing beyond 9 MHz. Some receivers have precision dial calibrations that are expressed in kilohertz (kHz). Megahertz frequency callouts can easily be converted to kilohertz by multiplying by 1000. Thus, 9 MHz becomes 9000 kHz; 15 MHz becomes 15,000 kHz; and 9.100 MHz becomes 9100 kHz.

Some receivers may use the terms kilocycles (kc) or megacycles (mc) instead of kilohertz and megahertz. The terms are synonymous; that is, "kilocycles" is the same as "kilohertz" and "megacycles" is the same as "megahertz." Formerly, the term "cycles per second" was used to designate frequencies. The terms kilocycles and megacycles (actually kilocycles per second and megacycles per second) were used to designate 1000 and 1,000,000 cycles per second. The newer term, hertz, was adopted partially because of the fact that the "per second" portion of the previous designation was often omitted (though without the time element the term is meaningless) and partially to honor Heinrich Hertz, considered by many as the father of radio. The term **hertz** (Hz) means **cycles per second**; thus, the time is included as part of the term. Likewise, kilohertz means 1000 cycles per second (1000 Hz) and megahertz means 1,000,000 cycles per second (1,000,000 Hz or 1000 kHz).

By international regulations, entered into by most countries, certain groups or "bands" of radio frequencies have been set aside in the high-frequency radio spectrum for international shortwave broadcasting. Most (but not all) of the world's shortwave broadcasting stations operate within these bands. Some, however, operate outside these bands, usually adjacent to a particular band within 100 or 200 kHz.

Occasionally you might hear that a particular station is operating in the "16-meter band," or the "41-meter band," etc. Table 1 contains a listing of international shortwave broadcasting bands and the frequencies in MHz of each. However, as mentioned previously, some stations do operate outside of the frequencies listed.

Table 1. International Shortwave Broadcasting Bands

Frequency (MHz)	Band (Meters)
2.300–2.498	120
3.200–3.400	90
3.900–4.000	75
4.750–5.060	60
5.950–6.200	49
7.100–7.300	41
9.500–9.775	31
11.700–11.975	25
15.100–15.450	19
17.700–17.900	16
21.450–21.750	13
25.600–26.100	11

Call Letters

Any listener of conventional radio and television is aware of the call letters assigned to transmitting stations. For example, KFI Los Angeles, California; KOA Denver, Colorado; WLS Chicago, Illinois; WNBC New York, New York—to name but a few. To a lesser degree this practice has also been carried over to the licensing of shortwave stations. Some (but not all) countries have assigned call letters to their high-frequency stations; however, the call letters are rarely used for station identification. Usually the announcer of a foreign shortwave station will merely say, "This is Radio Japan," "This is RSA, Radio South Africa," or "This is the Voice of America." Call letters listed in this book are for the convenience of the user, and are given only when available.

Power

Power, as listed in Section 1 of this book, refers to the radio-frequency power as radiated by the antenna of the shortwave station, and is expressed in kilowatts (kW). Most international shortwave stations use transmitting equipment with a radiation power of 50 kW or more to ride through the interference and atmospheric noise. This high power does not mean that stations of 5 kW or less cannot be heard. Quite the contrary—amateur shortwave operators have repeatedly disproved this by conversing with fellow "hams" all over the world using considerably less than 1 kW of power. The unpredictableness of shortwave listening is what makes the hobby interesting and the end result more rewarding.

Wave Propagation

Two types of radio-frequency waves are emitted from a shortwave transmitting antenna—the ground wave and the sky wave. The ground wave is of no significance for shortwave reception. The sky wave, however, on leaving the transmitting antenna travels upward at various angles above the surface of the earth. It would simply continue out into space were it not bent sufficiently to bring it back to the earth. The medium that causes such bending is the ionosphere, a region in the upper atmosphere where free ions and electrons exist in sufficient quantity to cause a change in the refractive index. Ultraviolet radiation from the sun is considered to be responsible for the ionization. For a given intensity of ionization, the amount of refraction becomes less as the frequency of the wave becomes higher. The bending is smaller, therefore, at high frequencies than it is at low frequencies. If the frequency is raised to a high enough value, the bending eventually will become too slight to bring the wave back to earth. At frequencies beyond this point, long-distance shortwave communication becomes impossible.

Because an increase in ionization causes an increase in the maximum frequency that can be bent sufficiently for long-distance communication, it can be seen that slight variations in sun radiation caused by sunspots, solar flares, and other solar disturbances can affect shortwave signal reception. At times, ionospheric conditions may cause a temporary "signal blackout" from some areas of the earth. Therefore, even though a station might be listed as being "on the air" for a particular time period, ionospheric conditions may prevent the signal from being heard.

Time Zones and Local Time

The United States is divided into seven standard time zones, designated as Eastern, Central, Mountain, Pacific, Yukon, Alaska/Hawaii, and Bering. These are set forth in the Uniform Time Act of 1966. The Canadian provinces occupy the first five of these seven zones, plus the Atlantic time zone on the east. In addition, Newfoundland and Labrador advance the clock one-half hour ahead of the Atlantic time. The various time zones are shown on the map in Fig. 1. Each time zone is approximately 15 degrees of longitude in width, and all places within a given zone use the time reckoned from the transit of the sun across the Standard Time Meridian of that zone. The time for each zone, starting with the Atlantic time zone and moving westward, is basically reckoned from the 60th, 75th, 90th, 105th, 120th, 135th, 150th, and 165th meridian west of Greenwich, England (prime meridian). The actual division line separating the various time zones wanders somewhat from these meridians to conform with local geographic areas and local convenience.

The time of all events contained in this book is given in Eastern Standard Time (EST). To obtain the **local** time of the event when the user lives in the Atlantic time zone, one hour must be added to the time shown. If he lives in the Central time zone, one hour must be subtracted from the listed time. For local time in the Mountain time zone, two hours must be subtracted, etc. Fig. 1 shows the number of hours to add or subtract to obtain local time.

The time at Greenwich, England, is designated as Greenwich Mean Time (GMT) or Universal Time (UT). This time is often used in international operations to avoid the confusion that can result in converting to local time. The time in the Eastern time zone is 5 hours slower than Greenwich Mean Time; that is, when it is 12 noon EST, it is 5 pm GMT.

The standard time differences for principal cities of the United States and Canada are listed in Table 2. All times listed are based on 12:00 noon EST. Table 3 gives a handy conversion from EST to the various other time zones. This table can be used in two ways. For example, if it is desired to convert a time listed in Section 1, 3, or 4 of this book to local time, first locate the listed time in the first (EST) column. Then read the local time directly opposite this time in the column for your time zone. Conversely, if you desire to know what stations might be broadcasting at a given local time, you can convert your local time to EST by locating the local time under the column for your local time zone and reading the EST from the left-hand column directly opposite it. For example, if it is 8 pm Pacific Daylight Time in your area and you want to know the Eastern Standard Time, first

Table 2. Standard Time Differences

At 12 o'clock noon Eastern Standard Time, the standard time in U.S.A. and Canadian cities is as follows:

City	Time	City	Time
Akron, Ohio	12:00 Noon	Las Vegas, Nev.	9:00 am
Albuquerque, N. M.	10:00 am	Louisville, Ky.	12:00 Noon
Anchorage, Alaska	7:00 am	Memphis, Tenn.	11:00 am
Atlanta, Ga.	12:00 Noon	Miami, Fla.	12:00 Noon
Austin, Tex.	11:00 am	Milwaukee, Wis.	11:00 am
Baltimore, Md.	12:00 Noon	Minneapolis, Minn.	11:00 am
Birmingham, Ala.	11:00 am	Mobile, Ala.	11:00 am
Bismark, N. Dak.	11:00 am	Montreal, Que., Canada	12:00 Noon
Boise, Idaho	10:00 am	Nashville, Tenn.	11:00 am
Boston, Mass.	12:00 Noon	Newark, N. J.	12:00 Noon
Buffalo, N. Y.	12:00 Noon	New Haven, Conn.	12:00 Noon
Butte, Mont.	10:00 am	New Orleans, La.	11:00 am
Charleston, S. C.	12:00 Noon	New York, N. Y.	12:00 Noon
Charlotte, N. C.	12:00 Noon	Nome, Alaska	6:00 am
Chattanooga, Tenn.	12:00 Noon	Norfolk, Va.	12:00 Noon
Cheyenne, Wyo.	10:00 am	Oklahoma City, Okla.	11:00 am
Chicago, Ill.	11:00 am	Omaha, Nebr.	11:00 am
Cincinnati, Ohio	12:00 Noon	Ottawa, Onto., Canada	12:00 Noon
Cleveland, Ohio	12:00 Noon	Peoria, Ill.	11:00 am
Colorado Springs, Colo.	10:00 am	Philadelphia, Pa.	12:00 Noon
Columbus, Ohio	12:00 Noon	Phoenix, Ariz.	10:00 am
Dallas, Tex.	11:00 am	Pierre, S. Dak.	11:00 am
Dayton, Ohio	12:00 Noon	Pittsburgh, Pa.	12:00 Noon
Denver, Colo.	10:00 am	Portland, Me.	12:00 Noon
Des Moines, Ia.	11:00 am	Portland, Ore.	9:00 am
Detroit, Mich.	12:00 Noon	Providence, R.I.	12:00 Noon
Duluth, Minn.	11:00 am	Quebec, Que., Canada	12:00 Noon
Dutch Harbor, Alaska	6:00 am	Reno, Nev.	9:00 am
Edmonton, Alta., Canada	10:00 am	Richmond, Va.	12:00 Noon
El Paso, Tex.	11:00 am	Rochester, N. Y.	12:00 Noon
Erie, Pa.	12:00 Noon	Sacramento, Calif.	9:00 am
Evansville, Ind.	11:00 am	St. Louis, Mo.	11:00 am
Fairbanks, Alaska	7:00 am	St. Paul, Minn.	11:00 am
Flint, Mich.	12:00 Noon	Salt Lake City, Utah	10:00 am
Fort Wayne, Ind.	12:00 Noon	San Antonio, Tex.	11:00 am
Fort Worth, Tex.	11:00 am	San Diego, Calif.	9:00 am
Frankfort, Ky.	12:00 Noon	San Francisco, Calif.	9:00 am
Galveston, Tex.	11:00 am	Santa Fe, N. M.	10:00 am
Gander, Nfld., Canada	1:30 pm	Savannah, Ga.	12:00 Noon
Grand Rapids, Mich.	12:00 Noon	Seattle, Wash.	9:00 am
Halifax, N. S., Canada	1:00 pm	Shreveport, La.	11:00 am
Hartford, Conn.	12:00 Noon	Sioux Falls, S. Dak.	11:00 am
Helena, Mont.	10:00 am	Spokane, Wash.	9:00 am
Hilo, Hawaii	7:00 am	Tacoma, Wash.	9:00 am
Honolulu, Hawaii	7:00 am	Tampa, Fla.	12:00 Noon
Houston, Tex.	11:00 am	Toledo, Ohio	12:00 Noon
Indianapolis, Ind.	12:00 Noon	Topeka, Kan.	11:00 am
Jacksonville, Fla.	12:00 Noon	Toronto, Ont., Canada	12:00 Noon
Juneau, Alaska	9:00 am	Tucson, Ariz.	10:00 am
Kansas City, Mo.	11:00 am	Tulsa, Okla.	11:00 am
Knoxville, Tenn.	12:00 Noon	Vancouver, B. C., Canada	9:00 am
Lexington, Ky.	12:00 Noon	Washington, D. C.	12:00 Noon
Lincoln, Nebr.	11:00 am	Wichita, Kan.	11:00 am
Little Rock, Ark.	11:00 am	Wilmington, Del.	12:00 Noon
Los Angeles, Calif.	9:00 am	Winnipeg, Man., Canada	11:00 am

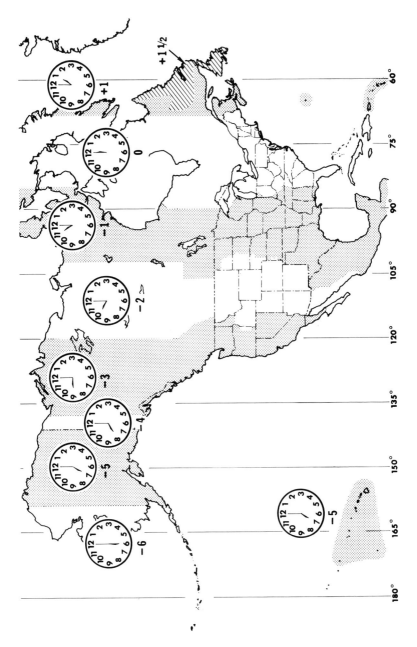

Fig. 1. North American Time Zones.

Table 3. Time Conversion Chart

EST CDT	AST EDT	CST MDT	MST PDT	PST	YST	AST HST	BST	GMT
Midnight	1 am	11 pm	10 pm	9 pm	8 pm	7 pm	6 pm	5 am
1 am	2 am	Midnight	11 pm	10 pm	9 pm	8 pm	7 pm	6 am
2 am	3 am	1 am	Midnight	11 pm	10 pm	9 pm	8 pm	7 am
3 am	4 am	2 am	1 am	Midnight	11 pm	10 pm	9 pm	8 am
4 am	5 am	3 am	2 am	1 am	Midnight	11 pm	10 pm	9 am
5 am	6 am	4 am	3 am	2 am	1 am	Midnight	11 pm	10 am
6 am	7 am	5 am	4 am	3 am	2 am	1 am	Midnight	11 am
7 am	8 am	6 am	5 am	4 am	3 am	2 am	1 am	Noon
8 am	9 am	7 am	6 am	5 am	4 am	3 am	2 am	1 pm
9 am	10 am	8 am	7 am	6 am	5 am	4 am	3 am	2 pm
10 am	11 am	9 am	8 am	7 am	6 am	5 am	4 am	3 pm
11 am	Noon	10 am	9 am	8 am	7 am	6 am	5 am	4 pm
Noon	1 pm	11 am	10 am	9 am	8 am	7 am	6 am	5 pm
1 pm	2 pm	Noon	11 am	10 am	9 am	8 am	7 am	6 pm
2 pm	3 pm	1 pm	Noon	11 am	10 am	9 am	8 am	7 pm
3 pm	4 pm	2 pm	1 pm	Noon	11 am	10 am	9 am	8 pm
4 pm	5 pm	3 pm	2 pm	1 pm	Noon	11 am	10 am	9 pm
5 pm	6 pm	4 pm	3 pm	2 pm	1 pm	Noon	11 am	10 pm
6 pm	7 pm	5 pm	4 pm	3 pm	2 pm	1 pm	Noon	11 pm
7 pm	8 pm	6 pm	5 pm	4 pm	3 pm	2 pm	1 pm	Midnight
8 pm	9 pm	7 pm	6 pm	5 pm	4 pm	3 pm	2 pm	1 am
9 pm	10 pm	8 pm	7 pm	6 pm	5 pm	4 pm	3 pm	2 am
10 pm	11 pm	9 pm	8 pm	7 pm	6 pm	5 pm	4 pm	3 am
11 pm	Midnight	10 pm	9 pm	8 pm	7 pm	6 pm	5 pm	4 am

locate 8 pm in the fourth column and then opposite this point in the first column, read the Eastern Standard Time (10 pm).

The Uniform Time Act of 1966 states that Daylight Savings Time (DST) will be observed from 2:00 am on the last Sunday in April to 2:00 am on the last Sunday in October. However, a few states have elected to exempt themselves from the observance of Daylight Savings Time. Daylight Savings Time is achieved by **advancing** the clocks one hour. For example, an event listed here for 9:00 pm EST would take place at 10:00 pm EDT.

Stations by Country and City

Location	Call Letters	Power (kW)	Freq (MHz)	Transmission Period (EST)
AFARS & ISSAS				
Djibouti		4.0	4.780	10:00 pm–Midnight
				Midnight–1:00 am
AFGHANISTAN				
Kabul		100	4.775	9:00 am–9:30 am
Kabul		100	9.530	1:00 pm–1:30 pm
Kabul		50	11.790	1:00 pm–1:30 pm
Kabul		100	15.265	1:00 pm–1:30 pm
Kabul		50	17.780	1:00 pm–1:30 pm
ALBANIA				
Tirana		50/500	6.195	8:30 pm–9:00 pm
				9:30 pm–10:00 pm
				10:30 pm–11:00 pm
Tirana		50/240	7.065	7:00 pm–7:30 pm
Tirana		50/500	7.300	8:30 pm–9:00 pm
				9:30 pm–10:00 pm
				10:30 pm–11:00 pm
Tirana		50/500	9.500	6:00 am–6:30 am
Tirana		50/500	9.780	7:00 pm–7:30 pm
ALGERIA				
Algiers		100	11.715	1:00 am–7:00 pm
Algiers		100	11.835	1:00 am–7:00 pm
Algiers		50	15.420	1:00 am–7:00 pm

Location	Call Letters	Power (kW)	Freq (MHz)	Transmission Period (EST)
ANGOLA				
Luanda	CR6RZ	100	5.960	Midnight–3:00 am
Luanda	CR6RZ	100	7.245	6:00 am–9:00 am
Luanda	CR6RZ	100	9.535	Midnight–7:00 pm
				Midnight–2:00 pm
Luanda	CR6RZ	10	11.875	3:00 am–2:00 pm
ARGENTINA				
Buenos Aires	LRY	20	9.760	1:00 pm–5:00 am
Buenos Aires	LRS	20	11.880	7:00 am–11:00 pm
AUSTRALIA				
Melbourne		250	6.055	4:00 pm–5:15 pm
Melbourne		50	9.570	1:45 am–2:45 am
Melbourne		100	9.580	7:15 am–8:15 am
Melbourne		50	11.710	7:15 am–8:15 am
Melbourne		100	11.765	1:45 am–2:45 am
Melbourne		50/250	11.840	6:30 pm–7:15 pm
Melbourne		100	15.320	8:00 pm–10:00 pm
Melbourne		50/250	17.715	7:30 pm–9:00 pm
				11:00 pm–2:00 am
Melbourne		50/100	17.795	8:00 pm–10:00 pm
Melbourne		250	21.485	1:15 am–3:00 am
Melbourne		50	21.740	8:00 pm–10:00 pm
AUSTRIA				
Vienna		100	6.155	11:00 pm–7:00 am
Vienna		100	7.245	1:00 am–3:00 am
Vienna		100	9.770	6:00 pm–7:00 pm
Vienna		100	15.145	7:00 pm–9:00 pm
BANGLADESH				
Dacca		50	9.690	9:30 am–10:00 am
Dacca		50	11.620	7:30 am–8:00 am
Dacca		50	11.650	12:15 pm–1:00 pm
Dacca		50	15.520	9:30 am–10:00 am
Dacca		50	17.935	7:30 am–8:00 am
BELGIUM				
Brussels		100	9.550	6:05 pm–6:15 pm
				7:50 pm–8:00 pm

Location	Call Letters	Power (kW)	Freq (MHz)	Transmission Period (EST)
BELGIUM (cont)				
Brussels		100	11.875	6:05 pm–6:15 pm
				7:50 pm–8:00 pm
BOLIVIA				
La Paz	CP58	10	6.005	4:00 am–11:30 pm
La Paz	CP6	10	9.555	6:00 am–8:00 am
				10:00 am–11:00 pm
BOTSWANA				
Gaberones		10	3.356	11:00 am–11:30 am
				11:15 pm–11:30 pm
Gaberones		10	4.845	11:00 am–11:30 am
				11:15 pm–11:30 pm
Gaberones		10	5.965	6:30 am–6:45 am
Gaberones		10	9.590	6:30 am–6.45 am
BRAZIL				
Brasilia	PRL	7.5	6.065	3:00 am–Midnight
Rio de Janeiro	ZYZ38	10	9.720	3:00 am–11:00 pm
Rio de Janeiro	ZYZ39	10	11.795	4:00 am–10:00 pm
Sao Paulo	ZYR63	10	6.125	3:00 am–Midnight
BRUNEI				
Tutong		10	7.215	7:00 am–9:30 am
				6:00 pm–7:30 pm
				10:00 pm–Midnight
BULGARIA				
Sofia		100	6.070	2:30 pm–3:00 pm
				4:30 pm–5:00 pm
Sofia		120	9.700	2:30 pm–3:00 pm
				4:30 pm–5:00 pm
				7:00 pm–8:00 pm
				11:00 pm–Midnight
BURMA				
Rangoon	XZK42	0.5	5.040	9:30 am–11:00 am
Rangoon	XZK4	0.5	7.120	9:00 pm–9:30 pm
Rangoon	XZK5	0.5	9.725	2:00 am–2:30 am

Location	Call Letters	Power (kW)	Freq (MHz)	Transmission Period (EST)
CAMBODIA				
Phnom Penh		50	4.907	6:30 pm–7:00 pm
				7:45 pm–8:00 pm
				12:45 am–1:00 am
				7:45 am– 8:00 am
CAMEROON				
Yaoundé		30	4.972	12:30 am–1:00 am
				8:30 am–9:45 am
Yaoundé		30	9.760	12:30 pm–1:45 pm
				8:00 am–9:30 am
				12:30 pm–1:45 pm
CANADA				
Montreal		50	5.970	3:30 am–4:30 am
Montreal		50	5.990	2:15 am–2:45 am
Montreal		250	9.625	2:15 am–2:45 am
				3:30 am–4:30 am
				7:15 am–8:15 am
				6:00 pm–6:30 pm
Montreal		250	11.945	6:00 pm–6:30 pm
Montreal		250	15.190	6:00 pm–6:30 pm
Montreal		250	15.325	2:15 am–2:45 am
				1:30 pm–2:15 pm
				4:15 pm–5:00 pm
Montreal		250	17.820	2:15 am–2:45 am
				1:30 pm–2:15 pm
CANARY ISLANDS				
Santa Cruz		50	11.800	3:00 pm–Midnight
Santa Cruz		50	15.365	3:00 pm–Midnight
CAPE VERDE ISLANDS				
Sao Vicente		1.0	3.910	10:00 am–11:00 am
				5:00 pm–8:00 pm
Sao Vicente		0.25	4.715	2:00 am–4:00 am
				6:00 am–8:00 am
				1:00 pm–3:00 pm
CENTRAL AFRICAN REPUBLIC				
Bangui		100	5.035	11:30 am–6:00 pm
Bangui		100	7.220	2:30 am–11:30 am

Location	Call Letters	Power (kW)	Freq (MHz)	Transmission Period (EST)
CHAD				
Fort Lamy		30	4.905	11:15 am–4:30 pm
				11:30 pm–1:30 am
Fort Lamy		30	7.120	7:30 am–11:15 am
Fort Lamy		4.0	9.615	7:30 am–11:15 am
CHINA (People's Rep. of)				
Peking		100	7.120	8:00 pm–11:00 pm
Peking		100	7.590	3:30 pm–5:30 pm
Peking		100	9.030	3:30 pm–5:30 pm
Peking		100	9.780	8:00 pm–9:00 pm
				10:00 pm–11:00 pm
Peking		100	11.650	3:30 pm–5:30 pm
Peking		100	11.675	7:00 pm–Midnight
Peking		100	15.060	7:00 pm–Midnight
Peking		100	15.095	10:00 pm–Midnight
Peking		100	15.385	10:00 pm–Midnight
Peking		100	15.510	7:00 am–9:00 am
Peking		100	17.715	8:00 pm–10:00 pm
Peking		100	17.735	10:00 pm–Midnight
CHINA (Rep. of) (Taiwan)				
Taipei	BED7	100	7.130	9:00 pm–11:00 pm
Taipei	BED73	100	9.685	1:00 pm–2:00 pm
Taipei	BED66	100	9.765	1:00 pm–2:00 pm
Taipei	BED69	100	11.825	1:00 pm–2:00 pm
				9:00 pm–11:00 pm
Taipei	BED60	100	15.125	1:00 pm–2:00 pm
				9:00 pm–11:00 pm
Taipei	BED49	100	15.345	9:00 pm–11:00 pm
Taipei	BED93	100	15.370	1:00 pm–2:00 pm
Taipei	BED39	100	17.720	1:00 pm–2:00 pm
				9:00 pm–11:00 pm
Taipei	BED95	100	17.780	9:00 pm–11:00 pm
Taipei	BED40	100	17.890	1:00 pm–2:00 pm
				9:00 pm–11:00 pm
COLOMBIA				
Bogotá	HJCT	50	6.030	6:00 am–Midnight
Bogotá	HJWT	25	6.183	6:00 am–Midnight
Bogotá	HJZO	25	11.825	6:00 am–Midnight

Location	Call Letters	Power (kW)	Freq (MHz)	Transmission Period (EST)
COMORO ISLANDS				
Moroni		4.0	3.331	10:30 am–2:30 pm
Moroni		4.0	7.260	4:00 am–6:00 am
				10:30 pm–Midnight
CONGO (Republic of)				
Brazzaville		50	15.190	12:15 am–12:30 am
				6:00 am–6:30 am
				3:15 pm–4:00 pm
Brazzaville		50	21.500	12:15 am–12:30 am
				6:00 am–6:30 am
COOK ISLANDS				
Rarotonga		1.0	5.045	11:30 am–1:30 pm
				11:30 pm–3:00 am
Rarotonga		1.0	9.695	5:45 pm–6:00 pm
COSTA RICA				
San José		1.0	6.075	24 Hours
CUBA				
Havana		50	9.525	1:30 am–3:00 am
Havana		50	9.760	8:00 pm–10:00 pm
Havana		100	11.760	10:30 pm–1:00 am
Havana		10	11.840	Midnight–1:00 am
				8:00 pm–10:00 pm
Havana		50	15.155	3:15 pm–4:45 pm
Havana		50	15.270	3:50 pm–4:50 pm
Havana		50	15.285	3:50 pm–4:50 pm
CZECHOSLOVAKIA				
Prague		100	5.930	11:30 am–Noon
				12:30 pm–1:30 pm
				2:00 pm–2:30 pm
				8:00 pm–9:00 pm
				10:00 pm–11:00 pm
Prague		200	6.055	2:00 am–3:00 am
				10:00 am–11:30 am
Prague		100	7.345	11:30 am–Noon
				12:30 pm–1:30 pm
				2:00 pm–2:30 pm

Location	Call Letters	Power (kW)	Freq (MHz)	Transmission Period (EST)
CZECHOSLOVAKIA (cont)				
Prague		100	7.345	8:00 pm–9:00 pm
				10:00 pm–11:00 pm
Prague		200	9.505	2:00 am–3:00 am
				10:00 am–10:30 am
Prague		100	9.540	8:00 pm–9:00 pm
				10:00 pm–11:00 pm
Prague		200	9.630	8:00 pm–9:00 pm
				10:00 pm–11:00 pm
Prague		100	11.990	8:00 pm–9:00 pm
				10:00 pm–11:00 pm
Prague		100	15.240	10:30 am–11:30 am
Prague		100	17.840	10:30 am–11:30 am
				12:30 pm–1:30 pm
Prague		100	21.700	2:00 am–3:00 am
DAHOMEY				
Cotonou		4.0	3.270	12:15 am–12:30 am
				1:15 am–1:30 am
				7:00 am–7:15 am
				11:45 am–1:00 pm
				2:00 pm–2:45 pm
Cotonou		30	4.870	12:15 am–12:30 am
				1:15 am–1:30 am
				7:00 am–7:15 am
				11:45 am–1:00 pm
				2:00 pm–2:45 pm
DOMINICAN REPUBLIC				
Santo Domingo	HISD	20	6.090	6:00 am–11:00 pm
Santo Domingo	HISD	50	9.505	6:00 am–11:00 pm
ECUADOR				
Quito	HCJB	100	5.960	2:15 am–5:00 am
Quito	HCJB	100	9.605	8:10 pm–Midnight
Quito	HCJB	50	9.745	2:15 am–6:00 am
Quito	HCJB	100	11.745	6:30 pm–7:00 pm
				8:00 pm–Midnight
Quito	HCJB	50	11.915	2:15 am–6:00 am
Quito	HCJB	100	15.115	Midnight–5:00 am
				7:45 am–11:30 am
				8:00 pm–Midnight

Location	Call Letters	Power (kW)	Freq (MHz)	Transmission Period (EST)
ECUADOR (cont)				
Quito	HCJB	100	15.300	2:00 pm–3:15 pm
Quito	HCJB	50	17.880	7:45 am–11:30 am
Quito	HCJB	50	21.460	7:45 am–11:30 am
				2:00 pm–3:15 pm
ELLICE ISLAND				
Ellice		5.0	3.230	2:00 am–3:00 am
				1:45 pm–4:00 pm
EL SALVADOR				
San Salvador	YSS	5.0	5.980	5:00 pm–Midnight
San Salvador	YSS	5.0	9.555	5:00 pm–Midnight
ETHIOPIA				
Addis Ababa		100	7.145	1:30 pm–2:15 pm
Addis Ababa		100	9.725	11:00 pm–11:25 pm
Addis Ababa		100	11.890	12:30 am–1:00 am
Addis Ababa		100	11.910	2:30 pm–3:15 pm
Addis Ababa		100	15.315	8:00 am–9:00 am
FIJI ISLANDS				
Suva		10	3.230	Midnight–4:15 am
				1:00 pm–4:15 pm
				10:45 pm–Midnight
Suva		10	6.005	4:15 pm–8:45 pm
FINLAND				
Helsinki	OIX2	100	9.585	9:00 pm–10:00 pm
Helsinki	OIX4	100	15.185	11:00 am–1:30 pm
FRANCE				
Paris		100	7.155	12:15 am–12:30 am
Paris		100	7.255	12:15 am–12:30 am
Paris		100	9.700	12:15 am–12:30 am
Paris		100	11.920	12:15 am–12:30 am
Paris		100	11.930	12:15 am–12:30 am
Paris		100	15.295	12:15 am–12:30 am
				6:00 am–6:15 am
				3:15 pm–4:00 pm
Paris		100	17.720	6:00 am–6:15 am
				3:15 pm–4:00 pm

Location	Call Letters	Power (kW)	Freq (MHz)	Transmission Period (EST)
FRANCE (cont)				
Paris		100	17.730	12:15 am–12:30 am
Paris		100	21.580	6:00 am–6:15 am
				3:15 pm–4:00 pm
GABON				
Libreville		4.0	3.300	11:30 pm–1:30 am
Libreville		100	4.777	11:30 pm–1:30 am
				11:30 am–6:00 pm
Libreville		100	7.270	1:30 am–11:30 am
GAMBIA				
Bathurst		4.0	4.820	1:30 am–3:00 am
				7:00 am–6:00 pm
GERMANY (Democratic Rep.)				
Berlin		100	5.955	8:00 pm–8:45 pm
				9:30 pm–10:15 pm
				10:30 pm–11:15 pm
Berlin		50	6.080	10:30 pm–11:15 pm
Berlin		50	6.165	10:30 pm–11:15 pm
Berlin		100	9.500	1:15 am–2:00 am
Berlin		50	9.730	8:00 pm–8:45 pm
				9:30 pm–10:15 pm
Berlin		50	15.115	7:00 am–7:45 am
Berlin		100	15.390	1:15 pm–2:00 pm
Berlin		100	17.880	10:30 am–11:15 am
Berlin		50	21.465	1:45 am–2:30 am
Berlin		50	21.475	1:15 pm–2:00 pm
Berlin		50	21.540	7:00 am–7:45 am
				9:00 am–9:45 am
Berlin		50	21.600	8:15 am–9:00 am
GERMANY (Federal Rep.)				
Cologne		250	6.075	Midnight–1:00 am
				10:45 pm–11:00 pm
				11:30 pm–Midnight
Cologne		250	6.145	Midnight–1:00 am
				11:30 pm–Midnight
Cologne		250	7.225	11:30 pm–Midnight

Location	Call Letters	Power (kW)	Freq (MHz)	Transmission Period (EST)
GERMANY (Federal Rep.) (cont)				
Cologne		250	7.235	10:45 pm–11:00 pm
Cologne		250	9.545	Midnight–1:00 am
				11:30 pm–Midnight
Cologne		250	9.565	11:30 pm–Midnight
Cologne		250	9.620	1:00 am–1:30 am
Cologne		250	9.690	12:45 am–1:30 am
Cologne		250	9.735	8:30 pm–10:00 pm
Cologne		250	11.785	1:00 am–1:30 am
Cologne		250	11.795	4:20 am–5:20 am
Cologne		250	11.905	12:45 am–1:30 am
Cologne		250	15.185	4:20 am–5:20 am
Cologne		250	15.320	1:00 am–1:30 am
Cologne		250	17.875	6:15 am–6:45 am
Cologne		250	21.540	6:15 am–6:45 am
GHANA				
Accra		100	6.130	9:00 am–4:15 pm
Accra		20	9.545	3:45 pm–5:15 pm
Accra		250	11.850	3:00 pm–4:00 pm
Accra		250	15.285	1:15 pm–2:00 pm
Accra		250	17.870	9:00 am–10:00 am
Accra		250	21.545	9:45 am–10:30 am
GREAT BRITAIN				
London		100	5.975	Noon–4:15 pm
London		100	6.050	Midnight–2:30 am
London		250	6.110	Midnight–2:00 am
				4:15 pm–Midnight
London		250	6.180	Midnight–2:30 am
				Noon–7:30 pm
London		250	7.120	1:00 pm–8:00 pm
London		100	7.130	5:00 pm–Midnight
London		250	9.410	Midnight–3:00 am
				10:00 am–6:00 pm
				11:00 pm–Midnight
London		100	9.510	7:30 pm–10:30 pm
London		100	11.780	4:15 pm–6:15 pm
London		250	12.095	4:00 am–1:00 pm
London		250	15.070	4:00 am–1:00 pm
London		250	15.260	10:00 am–10:15 am

Location	Call Letters	Power (kW)	Freq (MHz)	Transmission Period (EST)
GREAT BRITAIN (cont)				
London		100	17.790	1:00 am–2:30 am
				7:00 am–Noon
London		250	21.470	4:00 am–Noon
GREECE				
Athens		50	7.295	1:30 pm–2:00 pm
Athens		50	9.605	1:30 pm–2:00 pm
				2:45 pm–3:00 pm
				5:15 pm–6:00 pm
Athens		50	11.720	2:45 pm–3:00 pm
				5:15 pm–6:30 pm
GUATEMALA				
Guatemala City	TGWB	10	6.180	7:00 am–11:00 pm
Guatemala City	TGNB	5.0	9.505	6:00 am–11:00 pm
GUINEA (Portuguese)				
Bissau		10	5.045	1:00 am–7:00 pm
GUINEA (Republic)				
Conakry		18	7.125	1:00 am–3:30 am
				11:00 am–7:00 pm
Conakry		100	9.650	Noon–6:00 pm
Conakry		100	15.310	1:00 am–3:30 am
				11:00 am–7:00 pm
GUIANA (French)				
Cayenne		4.0	3.385	10:00 am–1:00 pm
				3:30 pm–8:00 pm
GUYANA				
Georgetown		2.0	3.265	4:15 am–9:45 pm
HAITI				
Cap Haitien	4VEH	0.35	9.770	6:00 am–10:30 pm
Cap Haitien	4VEJ	2.5	11.835	6:00 am–10:30 pm
HOLLAND				
Hilversum		300	6.020	4:30 am–6:00 am
				9:00 am–10:30 am
				1:30 pm–3:00 pm

Location	Call Letters	Power (kW)	Freq (MHz)	Transmission Period (EST)
HOLLAND (cont)				
Hilversum		100	6.085	1:30 pm–3:00 pm
Hilversum		100	7.290	4:30 pm–6:00 pm
Hilversum		100	9.715	Midnight–1:30 am
				3:00 am–4:30 am
				4:00 pm–6:00 pm
Hilversum		300	9.740	4:30 pm–6:00 pm
Hilversum		300	11.730	Midnight–3:00 am
				9:00 pm–10:20 pm
Hilversum		300	17.830	1:30 pm–3:00 pm
HONDURAS (British)				
Belize		1.0	3.300	5:00 pm–Midnight
HONDURAS (Republic)				
Tegucigalpa		5.0	4.820	5:00 am–Midnight
Tegucigalpa		5.0	6.050	8:00 am–Midnight
HUNGARY				
Budapest		15	7.220	2:30 pm–3:00 pm
				4:30 pm–5:00 pm
Budapest		100	9.833	Midnight–12:30 am
				2:30 pm–3:00 pm
				4:30 pm–5:00 pm
				8:00 pm–8:30 pm
				10:00 pm–10:30 pm
Budapest		100	11.910	Midnight–12:30 am
				2:30 pm–3:00 pm
				4:30 pm–5:00 pm
				8:00 pm–8:30 pm
				10:00 pm–10:30 pm
Budapest		15	15.165	Midnight–12:30 am
				8:00 pm–8:30 pm
				10:00 pm–10:30 pm
Budapest		3.0	17.840	8:00 pm–8:30 pm
Budapest		3.0	21.685	8:00 pm–8:30 pm
INDIA				
New Delhi		100	7.235	5:45 pm–7:00 pm
New Delhi		100	9.570	5:45 pm–7:00 pm
New Delhi		250	9.690	2:45 pm–3:45 pm
New Delhi		250	9.912	2:45 pm–5:30 pm

Location	Call Letters	Power (kW)	Freq (MHz)	Transmission Period (EST)
INDIA (cont)				
New Delhi		100	11.620	12:45 pm–2:45 pm
New Delhi		250	11.710	5:45 pm–7:00 pm
New Delhi		250	11.775	5:00 am–6:00 am
New Delhi		100	11.810	8:00 am–10:00 am
New Delhi		100	11.945	12:45 pm–2:45 pm
New Delhi		100	11.960	2:45 pm–3:45 pm
New Delhi		200	15.080	12:45 pm–2:45 pm
New Delhi		250	15.190	5:00 am–6:00 am
New Delhi		100	15.205	5:00 am–6:00 am
New Delhi		100	15.235	7:15 pm–8:15 pm
New Delhi		250	15.335	8:00 am–10:00 am
New Delhi		100	17.775	5:00 am–6:00 am
New Delhi		250	21.555	5:00 am–6:00 am
INDONESIA				
Jakarta		50	9.585	9:30 am–10:30 am
				2:00 pm–3:00 pm
				6:30 pm–7:00 pm
Jakarta		25	11.795	4:00 am–4:30 am
				9:30 am–10:30 am
				2:00 pm–3:00 pm
				6:30 pm–7:00 pm
IRAN				
Tehran		100	12.165	12:30 pm–3:30 pm
Tehran		250	15.084	12:30 pm–3:30 pm
IRAQ				
Baghdad		100	9.745	2:30 pm–3:30 pm
ISRAEL				
Jerusalem		50	9.009	4:30 pm–5:00 pm
Jerusalem		100	9.625	4:30 pm–5:00 pm
Jerusalem		7.5	9.725	4:30 pm–5:00 pm
ITALY				
Rome		100	5.990	5:00 pm–5:30 pm
Rome		100	7.235	3:30 pm–3:45 pm
Rome		100	7.275	2:30 pm–3:00 pm
Rome		100	9.575	3:30 pm–3:45 pm

Location	Call Letters	Power (kW)	Freq (MHz)	Transmission Period (EST)
ITALY (cont)				
Rome		100	9.575	5:00 pm–5:30 pm
				8:00 pm–8:30 pm
Rome		100	9.710	2:30 pm–3:00 pm
Rome		100	11.810	2:30 pm–3:00 pm
				3:00 pm–3:45 pm
				8:00 pm–8:30 pm
Rome		100	15.330	10:45 pm–11:15 pm
Rome		100	17.795	10:45 pm–11:15 pm
IVORY COAST				
Abidjan		10	7.215	1:00 am–3:00 am
				1:00 pm–7:00 pm
Abidjan		100	11.920	1:00 am–3:00 am
				1:00 pm–7:00 pm
JAPAN				
Tokyo		100	9.505	2:00 am–2:15 am
				3:00 am–3:15 am
				4:00 am–4:15 am
				5:00 am–5:15 am
				7:00 am–7:15 am
				8:00 am–8:15 am
				9:00 am–9:15 am
				11:00 am–11:15 am
				Noon–12:15 pm
				1:00 pm–1:15 pm
Tokyo		100	11.815	5:00 am–5:15 am
				7:00 am–7:15 am
				8:00 am–8:15 am
				9:00 am–9:15 am
				11:00 am–11:15 am
				Noon–12:15 pm
				1:00 pm–1:15 pm
				2:00 pm–2:15 pm
				3:00 pm–3:15 pm
				4:00 pm–4:15 pm
				5:00 pm–5:15 pm
				6:00 pm–6:15 pm
Tokyo		100	15.105	2:00 pm–2:15 pm
				3:00 pm–3:15 pm
				4:00 pm–4:15 pm

Location	Call Letters	Power (kW)	Freq (MHz)	Transmission Period (EST)
JAPAN (Cont)				
Tokyo		100	15.195	2:00 am–2:15 am
				3:00 am–3:15 am
				4:00 am–4:15 am
Tokyo		100	15.390	5:00 am–5:15 am
				7:00 am–7:15 am
				8:00 am–8:15 am
				9:00 am–9:15 am
				11:00 am–11:15 am
				Noon–12:15 pm
				1:00 pm–1:15 pm
				2:00 pm–2:15 pm
				3:00 pm–3:15 pm
				4:00 pm–4:15 pm
				5:00 pm–5:15 pm
				6:00 pm–6:15 pm
Tokyo		100	17.785	5:00 pm–5:15 pm
				6:00 pm–6:15 pm
				8:00 pm–8:15 pm
				9:00 pm–9:15 pm
				10:00 pm–10:15 pm
				11:00 pm–11:15 pm
Tokyo		100	17.855	Midnight–12:15 am
				1:00 am–1:15 am
				2:00 am–2:15 am
				3:00 am–3:15 am
				4:00 am–4:15 am
				8:00 pm–8:15 pm
				9:00 pm–9:15 pm
				10:00 pm–10:15 pm
				11:00 pm–11:15 pm
Tokyo		10	17.880	Midnight–12:15 am
				1:00 am–1:15 am
				8:00 pm–8:15 pm
				9:00 pm–9:15 pm
				10:00 pm–10:15 pm
				11:00 pm–11:15 pm
JORDAN				
Amman		100	7.155	5:00 am–8:30 am
Amman		100	9.560	9:00 am–Noon

Location	Call Letters	Power (kW)	Freq (MHz)	Transmission Period (EST)
KENYA				
Nairobi		100	4.915	Midnight–1:30 am
				4:00 am–6:00 am
				8:00 am–3:00 pm
				10:00 pm–Midnight
Nairobi		100	7.140	4:00 am–6:00 am
KOREA (Dem. People's Rep.)				
Pyongyang		50	6.540	3:00 am–4:00 am
				2:00 pm–3:00 pm
				9:00 pm–10:00 pm
Pyongyang		50	9.515	8:00 am–10:00 am
				2:00 pm–3:00 pm
Pyongyang		50	15.150	3:00 am–4:00 am
				6:00 am–7:00 am
				8:00 am–10:00 am
				9:00 pm–10:00 pm
KOREA (Republic of)				
Seoul	HLK53	10	6.135	300 am–3:30 am
				6:30 am–6:30 am
Seoul	HLK5	50	9.640	Midnight–12:30 am
				4:00 am–4:30 am
				6:00 am–6:30 am
				4:00 pm–4:30 pm
Seoul	HLK41	50	15.155	1:30 am–2:00 am
Seoul	HLK41	50	15.430	4:00 am–4:30 am
				8:30 am–9:00 am
				4:00 pm–4:30 pm
				10:00 pm–10:30 pm
KUWAIT				
Kuwait		250	9.595	11:00 am–1:00 pm
Kuwait		250	11.925	1:30 pm–4:00 pm
Kuwait		250	15.345	11:00 am–2:00 pm
Kuwait		250	17.750	Midnight–1:00 am
				11:00 pm–Midnight
LEBANON				
Beirut		100	11.705	1:15 pm–3:30 pm

Location	Call Letters	Power (kW)	Freq (MHz)	Transmission Period (EST)
LESOTHO				
Maseru		10	4.800	Midnight–2:00 am
				5:30 am–7:00 am
				10:00 am–3:00 pm
				11:00 pm–Midnight
LIBERIA				
Monrovia	ELWA	10	4.770	1:15 am–3:15 am
				11:15 am–2:00 pm
Monrovia	ELWA	50	11.940	2:15 pm–4:00 pm
Monrovia	ELWA	50	11.950	1:30 am–2:15 am
LUXEMBOURG				
Villa Louvigny		50	6.090	1:30 pm–9:00 pm
MALAGASY (Republic)				
Tananarive		100	17.730	8:30 am–9:30 am
MALAWI				
Zomba		10	3.380	Midnight–12:30 am
				3:00 am–4:00 am
				1:00 pm–5:00 pm
				11:00 pm–Midnight
Zomba		10	5.995	Midnight–12:30 am
				3:00 am–4:00 am
				1:00 pm–5:00 pm
				11:00 pm–Midnight
MALAYSIA				
Kuala Lumpur		10	9.660	12:30 am–1:30 am
				4:30 am–11:30 am
				5:30 pm–8:30 pm
Kuala Lumpur		100	11.900	12:30 am–4:00 am
				4:30 am–11:30 am
				5:30 pm–8:30 pm
Kuala Lumpur		50	15.280	12:30 am–4:00 am
				4:30 am–11:30 am
				5:30 pm–8:30 pm
MALI				
Bamako		100	17.725	10:00 am–12:30 pm

Location	Call Letters	Power (kW)	Freq (MHz)	Transmission Period (EST)
MAURITANIA				
Nouakchott		30	4.850	2:00 am–3:00 am
				Noon–6:00 pm
Nouakchott		4.0	7.245	7:00 am–9:00 am
Nouakchott		30	9.610	7:00 am–9:00 am
MAURITIUS				
Forest Side		10	4.850	8:00 am–1:30 pm
Forest Side		10	9.710	Midnight–8:00 am
				9:00 pm–Midnight
MEXICO				
Mexico City	XERMX	100	11.770	9:00 am–1:00 am
Mexico City	XERMX	100	17.835	9:00 am–9:00 pm
Mexico City	XERMX	100	21.705	9:00 am–10:00 pm
MONACO				
Monte Carlo		30	6.035	Midnight–8:00 pm
Monte Carlo		30	7.135	Midnight–8:00 pm
MONGOLIA				
Ulan Bator		50	9.540	5:00 pm–5:30 pm
Ulan Bator		25	11.860	5:00 pm–5:30 pm
Ulan Bator		50	15.445	7:20 am–7:40 am
Ulan Bator		50	17.780	7:20 am–7:40 am
MOROCCO				
Tangier		50	7.225	Noon–1:00 pm
Tangier		50	11.735	Noon–1:00 pm
MOZAMBIQUE				
Lourenco Marques		100	6.050	Midnight–11:00 am
				11:00 pm–Midnight
Lourenco Marques		10	9.620	2:00 am–9:00 am
Lourenco Marques		10	11.780	Midnight–2:00 pm
NEPAL				
Katmandu		100	7.165	2:20 am–4:30 am
Katmandu		100	9.595	2:00 am–4:00 am
				8:20 pm–11:00 pm

Location	Call Letters	Power (kW)	Freq (MHz)	Transmission Period (EST)
NETHERLANDS ANTILLES				
Bonaire		50	11.820	6:00 am–7:30 am
				7:30 pm–8:30 pm
Bonaire		200	15.255	7:30 am–10:00 am
NEW CALEDONIA				
Noumea		20	3.355	1:00 am–6:00 am
				2:00 pm–4:00 pm
				6:30 pm–9:00 pm
Noumea		20	7.170	1:00 am–6:00 am
				2:00 pm–4:00 pm
				6:30 pm–9:00 pm
NEW GUINEA				
Bougainville		2.0	3.322	2:00 am–7:00 am
Milne Bay		10	3.360	2:30 am–7:00 am
Port Moresby		10	9.520	5:15 pm–2:00 am
Port Moresby		10	11.880	8:00 pm–12:30 am
NEW ZEALAND				
Wellington		7.5	11.705	4:00 am–7:00 am
Wellington		7.5	11.780	1:00 am–4:00 am
				Noon–3:00 pm
Wellington		7.5	15.110	3:00 pm–Midnight
Wellington		7.5	17.770	3:00 pm–Midnight
NICARAGUA				
Managua	YNM	100	11.875	8:00 am–Midnight
NIGER				
Niamey		4.0	7.155	6:30 am–8:30 am
NIGERIA				
Lagos		100	7.275	12:30 am–2:30 am
				10:30 am–Noon
				1:00 pm–2:30 pm
Lagos		100	11.925	1:00 pm–2:30 pm
Lagos		100	15.120	12:30 am–2:30 am
				10:30 am–Noon
Lagos		100	15.185	12:30 am–2:30 am
				10:30 am–Noon
				1:00 pm–2:30 pm

Location	Call Letters	Power (kW)	Freq (MHz)	Transmission Period (EST)
NORWAY				
Oslo		120	9.550	8:00 pm–8:30 pm
Oslo		120	9.645	Midnight–12:30 am
				6:00 pm–6:30 pm
				8:00 pm–8:30 pm
Oslo		120	11.850	2:00 am–2:30 am
				Noon–12:30 pm
				2:00 pm–2:30 pm
				4:00 pm–4:30 pm
Oslo		120	15.175	2:00 am–2:30 am
Oslo		120	21.655	2:00 am–2:30 am
				6:00 am–6:30 am
				10:00 am–10:30 am
				2:00 pm–2:30 pm
PAKISTAN				
Karachi		100	7.095	1:45 pm–4:30 pm
Karachi		100	7.265	7:15 pm–8:00 pm
Karachi		100	9.460	1:45 pm–4:30 pm
Karachi		100	21.590	Midnight–5:15 am
				8:00 am–9:00 am
				11:45 pm–Midnight
PANAMA				
David		1.0	6.045	6:00 am–Midnight
PARAGUAY				
Asunción	ZPA6	3.0	5.975	7:00 am–1:00 pm
				4:00 pm–10:00 pm
Asunción	ZPA5	5.0	11.945	4:30 am–Noon
Asunción	ZPA7	3.0	15.210	4:30 am–10:00 pm
PHILIPPINES				
Manila	DZF2	50	11.920	6:45 am–9:00 am
Manila	DZH9	50	15.235	6:30 pm–7:00 pm
Manila	DZH9	50	15.300	8:30 pm–9:00 pm
Manila	DZF8	50	15.440	8:00 pm–1:30 am
Manila	DZI6	50	17.810	9:00 pm–1:00 am
Manila	DZI8	50	21.515	Midnight–1:30 am
				7:00 pm–Midnight

Location	Call Letters	Power (kW)	Freq (MHz)	Transmission Period (EST)
POLAND				
Warsaw		100	6.035	9:00 pm–10:30 pm
Warsaw		100	6.135	9:00 pm–10:30 pm
Warsaw		100	7.285	9:00 pm–10:30 pm
Warsaw		100	11.815	9:00 pm–10:30 pm
Warsaw		100	11.840	9:00 pm–10:30 pm
Warsaw		100	15.120	9:00 pm–10:30 pm
Warsaw		100	15.275	9:00 pm–10:30 pm
PORTUGAL				
Lisbon		100	6.025	3:45 pm–4:30 pm
				10:45 pm–11:30 pm
Lisbon		100	11.935	1:15 pm–2:15 pm
				9:00 pm–9:45 pm
				10:45 pm–11:30 pm
Lisbon		100	17.880	2:30 am–4:00 am
Lisbon		100	17.895	8:45 am–9:30 am
Lisbon		100	21.495	2:30 am–4:00 am
				8:45 am–9:30 am
				1:15 pm–2:15 pm
REUNION				
St. Denis		4.0	4.807	Midnight–1:45 pm
				9:30 pm–Midnight
RHODESIA				
Salisbury		100	3.396	Midnight–1:00 am
				10:45 am–4:00 pm
				11:00 pm–Midnight
Salisbury		100	7.285	1:00 am–10:45 am
ROMANIA				
Bucharest		120	5.980	8:30 pm–9:30 pm
				10:00 pm–Midnight
Bucharest		120	6.150	2:30 pm–3:00 pm
				8:30 pm–9:30 pm
				10:00 pm–10:30 pm
				11:30 pm–Midnight
Bucharest		120	6.190	4:00 pm–4:30 pm
				8:30 pm–9:30 pm
				10:00 pm–10:30 pm
				11:30 pm–Midnight

Location	Call Letters	Power (kW)	Freq (MHz)	Transmission Period (EST)
ROMANIA (cont)				
Bucharest		120	7.195	2:30 pm–3:00 pm
Bucharest		120	7.225	4:00 pm–4:30 pm
Bucharest		120	9.510	8:30 pm–9:30 pm
				10:00 pm–10:30 pm
				11:30 pm–Midnight
Bucharest		120	9.570	8:30 pm–9:30 pm
				10:00 pm–10:30 pm
				11:30 pm–Midnight
Bucharest		120	9.690	8:30 pm–9:30 pm
				10:00 pm–10:30 pm
				11:30 pm–Midnight
Bucharest		120	11.920	6:00 am–6:30 am
				10:00 am–10:30 am
Bucharest		120	11.940	8:30 pm–9:30 pm
				10:00 pm–10:30 pm
				11:30 pm–Midnight
Bucharest		120	15.250	1:45 am–2:15 am
				6:00 am–6:30 am
				8:00 am–8:30 am
				10:00 am–10:30 am
Bucharest		120	17.710	8:00 am–8:30 am
Bucharest		120	17.850	1:45 am–2:15 am
				6:00 am–6:30 am
RWANDA				
Kigali		50	6.055	Midnight–1:00 am
				4:00 am–7:00 am
				8:30 am–4:00 pm
				10:30 pm–Midnight
SARAWAK				
Kuching		25	7.160	6:00 am–6:30 am
				9:00 am–9:45 am
				6:30 pm–8:00 pm
SAUDI ARABIA				
Mecca		50	11.855	Midnight–1:00 am
				6:00 am–7:00 am
				Noon–3:00 pm
				11:30 pm–Midnight

Location	Call Letters	Power (kW)	Freq (MHz)	Transmission Period (EST)
SENEGAL				
Dakar		25	4.890	1:15 pm–1:45 pm
Dakar		100	11.895	1:15 pm–1:45 pm
SEYCHELLES				
Victoria		50	11.920	9:45 pm–10:15 pm
Victoria		50	11.930	8:30 am–9:30 am
				10:30 am–11:45 am
Victoria		50	15.185	9:45 pm–10:15 pm
Victoria		50	15.270	8:30 am–9:30 am
				10:30 am–11:45 am
SINGAPORE				
Singapore		10	5.052	Midnight–11:30 am
				5:30 pm–Midnight
Singapore		50	11.940	Midnight–11:30 am
				5:30 pm–Midnight
SOLOMON ISLANDS				
Honiara		5.0	9.775	2:00 am–6:15 am
				2:00 pm–7:00 pm
				8:00 pm–10:00 pm
SOMALIA				
Mogadishu		50	9.588	12:30 pm–1:00 pm
SOUTH AFRICA				
Johannesburg		100	6.080	6:30 pm–10:30 pm
Johannesburg		100	7.270	Midnight–2:00 am
				11:15 pm–Midnight
Johannesburg		100	9.525	Midnight–12:15 am
				5:15 pm–6:15 pm
				11:15 pm–Midnight
Johannesburg		100	9.560	6:30 pm–10:30 pm
Johannesburg		100	9.695	2:00 pm–3:00 pm
				5:15 pm–10:30 pm
Johannesburg		100	11.900	Midnight–12:15 am
				1:30 am–1:45 am
				6:00 am –10:00 am
				11:00 am–Noon
				5:15 pm–6:15 pm
Johannesburg		100	11.970	5:15 pm–10:30 pm

Location	Call Letters	Power (kW)	Freq (MHz)	Transmission Period (EST)
SOUTH AFRICA (cont)				
Johannesburg		100	15.155	1:00 pm–2:00 pm
Johannesburg		100	15.175	2:00 pm–3:00 pm
Johannesburg		100	15.220	Midnight–1:45 am
				5:00 am–Noon
				11:15 pm–Midnight
Johannesburg		100	17.805	Midnight–12:15 am
				1:30 am–1:45 am
				11:15 pm–Midnight
Johannesburg		100	17.815	3:00 am–4:00 am
Johannesburg		100	17.820	5:00 am–6:00 am
Johannesburg		100	21.480	1:00 pm–2:00 pm
Johannesburg		100	21.520	5:00 am–6:00 am
Johannesburg		100	21.535	1:30 am–1:45 am
				4:30 am–Noon
Johannesburg		100	21.545	3:00 am–4:00 am
Johannesburg		100	25.790	4:30 am–4:45 am
SPAIN				
Madrid		100	6.140	8:00 pm–11:00 pm
SRI LANKA (Ceylon)				
Colombo		10	4.968	7:30 am–Noon
Colombo		10	5.020	6:30 am–Noon
Colombo		10	6.075	8:00 pm–9:00 pm
Colombo		100	9.720	8:00 pm–11:30 pm
Colombo		100	15.120	8:00 pm–9:30 pm
Colombo		35	17.830	6:00 am–7:00 am
SUDAN				
Omdurman		20	6.150	12:15 pm–1:00 pm
Omdurman		120	7.200	12:15 pm–1:00 pm
Omdurman		50	9.505	12:15 pm–1:00 pm
SWEDEN				
Stockholm		100	6.065	7:30 am–8:00 am
				11:00 am–11:30 am
				3:45 pm–4:15 pm
Stockholm		100	6.175	7:30 pm–8:00 pm
				9:00 pm–9:30 pm
Stockholm		100	9.585	12:15 am–12:45 am

Location	Call Letters	Power (kW)	Freq (MHz)	Transmission Period (EST)
SWEDEN (cont)				
Stockholm		100	9.630	6:00 am–6:30 am
				7:30 am–8:00 am
Stockholm		100	9.745	3:45 pm–4:15 pm
Stockholm		100	11.705	10:30 pm–11:00 pm
Stockholm		100	11.930	11:00 am–11:30 am
Stockholm		100	11.950	1:30 pm–2:00 pm
Stockholm		100	15.240	9:00 am–9:30 am
				1:30 pm–2:00 pm
Stockholm		100	17.840	12:15 am–12:45 am
Stockholm		100	21.505	9:00 am–9:30 am
Stockholm		100	21.690	6:00 am–6:30 am
SWITZERLAND				
Berne		250	6.120	8:30 pm–9:00 pm
				11:00 pm–11:30 pm
Berne		250	9.535	8:30 pm–9:00 pm
				11:00 pm–11:30 pm
Berne		250	9.590	2:00 am–2:30 am
				4:00 pm–4:45 pm
Berne		250	9.750	8:30 pm–9:00 pm
Berne		250	11.715	8:30 pm–9:00 pm
Berne		250	11.720	4:00 pm–4:45 pm
Berne		250	11.775	2:00 am–2:30 am
Berne		100	11.865	3:45 am–4:15 am
				10:15 am–10:45 am
				4:00 pm–4:45 pm
Berne		100	15.305	3:45 am–4:15 am
				8:15 am–8:45 am
				10:15 am–10:45 am
				4:00 pm–4:45 pm
Berne		100	15.430	6:00 am–6:30 am
Berne		100	17.795	6:00 am–6:30 am
Berne		100	17.830	10:15 am–10:45 am
Berne		100	17.845	8:15 am–8:45 am
Berne		100	21.520	6:00 am–6:30 am
				8:15 am–8:45 am
Berne		100	21.585	6:00 am–6:30 am
Berne		100	21.605	8:15 am–8:45 am

TAIWAN (See China)

Location	Call Letters	Power (kW)	Freq (MHz)	Transmission Period (EST)
TANZANIA				
Dar es Salaam		10	4.785	11:00 am–1:30 pm
				10:45 pm–11:45 pm
Dar es Salaam		100	6.105	11:00 am–1:15 pm
				10:45 pm–11:45 pm
Dar es Salaam		10	7.280	4:00 am–5:30 am
Dar es Salaam		100	9.750	4:00 am–5:30 am
Dar es Salaam		50	15.435	1:30 pm–2:45 pm
THAILAND				
Bangkok		5.0	7.115	Midnight–12:15 am
				5:30 am–6:30 am
				11:15 pm–Midnight
Bangkok		100	11.910	Midnight–12:15 am
				5:30 am–6:30 am
				11:15 pm–Midnight
TOGO				
Lomé		100	5.047	2:45 pm–3:00 pm
Lomé		100	7.265	7:45 am–8:00 am
				2:45 pm–3:00 pm
TURKEY				
Ankara		250	9.515	5:00 pm–5:30 pm
Ankara		250	15.195	5:00 pm–5:30 pm
Ankara		250	17.820	8:30 am–9:00 am
TURKS & CAICOS ISLANDS				
Grand Turk	VSI8	0.75	4.788	2:30 pm–3:30 pm
UGANDA				
Kampala		20	7.110	4:00 am–4:15 am
				9:00 am–9:30 am
Kampala		20	7.195	4:00 am–4:15 am
				9:00 am–9:30 am
UNITED ARAB REPUBLIC (Egypt)				
Cairo		250	9.475	9:00 pm–10:00 pm
Cairo		250	9.850	4:45 pm–6:00 pm
Cairo		250	17.655	12:30 pm–1:45 pm
Cairo		250	17.725	3:30 pm–5:00 pm
Cairo		250	17.920	8:15 am–9:30 am

Location	Call Letters	Power (kW)	Freq (MHz)	Transmission Period (EST)
UNITED NATIONS				
New York, N. Y.		100	5.955	3:45 am–4:00 am
New York, N. Y.		100	11.850	3:45 am–4:00 am
New York, N. Y.		100	15.155	9:30 pm–9:45 pm
New York, N. Y.		100	17.830	9:30 pm–9:45 pm
UNITED STATES OF AMERICA				
Belmont, Calif.	KGEI	250	9.670	11:30 pm–Midnight
Belmont, Calif.	KGEI	250	15.280	11:30 pm–Midnight
N. Y., N. Y.	WNYW	100	9.690	3:00 pm–6:00 pm
N. Y., N. Y.	WNYW	100	11.890	3:00 pm–5:15 pm
N. Y., N. Y.	WNYW	100	15.215	2:00 pm–2:45 pm
N. Y., N. Y.	WNYW	100	17.845	Noon–2:45 pm
N. Y., N. Y.	WNYW	100	21.525	Noon–1:45 pm
Wash., D. C.	(VOA)	100	3.980	Midnight–2:30 am
				11:00 am–7:00 pm
				10:00 pm–Midnight
Wash., D. C.	(VOA)	100	5.955	10:00 pm–2:30 am
Wash., D. C.	(VOA)	100	5.965	10:00 pm–1:00 am
Wash., D. C.	(VOA)	100	5.995	10:00 pm–2:30 am
Wash., D. C.	(VOA)	100	6.015	11:00 pm–Midnight
Wash., D. C.	(VOA)	100	6.040	11:00 am–7:00 pm
Wash., D. C.	(VOA)	100	6.160	10:00 pm–2:30 am
				1:00 pm–7:30 pm
Wash., D. C.	(VOA)	100	7.195	11:00 pm–2:30 am
Wash., D. C.	(VOA)	100	7.200	10:00 pm–2:30 am
Wash., D. C.	(VOA)	100	7.280	11:00 pm–2:30 am
Wash., D. C.	(VOA)	100	7.285	1:00 pm–7:00 pm
Wash., D. C.	(VOA)	100	9.530	11:00 pm–2:30 am
Wash., D. C.	(VOA)	100	9.635	10:00 pm–2:30 am
Wash., D. C.	(VOA)	100	9.740	10:00 pm–2:30 am
Wash., D. C.	(VOA)	100	9.750	11:00 pm–2:30 am
Wash., D. C.	(VOA)	100	9.760	1:00 pm–7:00 pm
Wash., D. C.	(VOA)	100	11.740	1:00 am–2:30 am
Wash., D. C.	(VOA)	100	11.760	10:00 pm–2:30 am
				5:00 pm–7:00 pm
Wash., D. C.	(VOA)	100	11.790	9:00 am–6:15 pm
Wash., D. C.	(VOA)	100	11.830	1:00 am–2:00 am
Wash., D. C.	(VOA)	100	11.970	1:00 am–2:00 am
Wash., D. C.	(VOA)	100	15.205	11:00 am–7:00 pm
Wash., D. C.	(VOA)	100	17.785	11:00 am–1:00 pm

Location	Call Letters	Power (kW)	Freq (MHz)	Transmission Period (EST)
UNITED STATES OF AMERICA (cont)				
Wash., D. C.	(VOA)	100	21.485	9:00 am–6:15 pm
Wash., D. C.	(VOA)	100	21.650	9:00 am–6:15 pm
UPPER VOLTA				
Ouagadougou		4.0	9.515	1:00 am–3:00 am
				7:00 am–9:00 am
				Noon–6:00 pm
U. S. S. R.				
Moscow		120	9.660	Midnight–12:30 am
				5:00 pm–5:30 pm
				6:00 pm–6:30 pm
				7:00 pm–7:30 pm
				8:00 pm–10:30 pm
				11:00 pm–Midnight
Moscow		120	9.668	Midnight–12:30 am
				5:00 pm–5:30 pm
				6:00 pm–6:30 pm
				7:00 pm–7:30 pm
				8:00 pm–10:30 pm
				11:00 pm–Midnight
Moscow		120	11.850	Midnight–12:30 am
				5:00 pm–5:30 pm
				6:00 pm–6:30 pm
				7:00 pm–7:30 pm
				8:00 pm–10:30 pm
				11:00 pm–Midnight
Moscow		120	11.870	Midnight–12:30 am
				5:00 pm–5:30 pm
				6:00 pm–6:30 pm
				7:00 pm–7:30 pm
				8:00 pm–10:30 pm
				11:00 pm–Midnight
Moscow		120	11.900	Midnight–12:30 am
				5:00 pm–5:30 pm
				6:00 pm–6:30 pm
				7:00 pm–7:30 pm
				8:00 pm–10:30 pm
				11:00 pm–Midnight
Moscow		120	11.920	Midnight–12:30 am
				5:00 pm–5:30 pm

Location	Call Letters	Power (kW)	Freq (MHz)	Transmission Period (EST)
U.S.S.R. (cont)				
Moscow		120	11.920	6:00 pm–6:30 pm
				7:00 pm–7:30 pm
				8:00 pm–10:30 pm
				11:00 pm–Midnight
Moscow		120	15.210	Midnight–12:30 am
				5:00 pm–5:30 pm
				6:00 pm–6:30 pm
				7:00 pm–7:30 pm
				8:00 pm–10:30 pm
				11:00 pm–Midnight
Moscow		120	15.230	11:00 am–Noon
Moscow		120	15.295	10:00 am–11:00 am
Moscow		120	17.730	Midnight–12:30 am
				5:00 pm–5:30 pm
				6:00 pm–6:30 pm
				7:00 pm–7:30 pm
				8:00 pm–10:30 pm
				11:00 pm–Midnight
Moscow		120	17.775	11:00 am–Noon
Moscow		120	17.865	Midnight–1:00 am
VATICAN STATE				
Vatican City		100	6.190	3:00 pm–4:00 pm
Vatican City		100	7.250	2:30 am–3:00 am
				3:00 pm–4:00 pm
Vatican City		100	9.615	7:50 am–8:15 am
Vatican City		100	9.645	10:00 am–10:30 am
				3:30 pm–4:00 pm
Vatican City		100	11.705	11:20 am–3:00 pm
Vatican City		100	11.725	7:50 am–8:15 am
Vatican City		100	11.740	2:30 am–3:30 am
				10:00 am–10:30 am
				3:45 pm–4:00 pm
Vatican City		100	15.120	2:30 am–3:30 am
				10:00 am–10:30 am
				3:45 pm–4:00 pm
Vatican City		100	15.155	7:50 am–8:15 am
Vatican City		100	15.210	11:30 am–3:00 pm
Vatican City		100	17.840	5:00 am–7:00 am
Vatican City		100	21.485	5:00 am–7:00 am

Location	Call Letters	Power (kW)	Freq (MHz)	Transmission Period (EST)
VIETNAM (Democratic Rep.)				
Hanoi		50	7.470	6:30 am–7:00 am
				8:30 am–9:00 am
				9:30 am–10:00 am
				10:30 am–11:00 am
Hanoi		50	10.010	8:30 am–9:00 am
				9:30 am–10:00 am
				10:30 am–11:00 am
				6:30 pm–7:00 pm
Hanoi		50	10.040	Midnight–12:30 am
				3:30 am–4:00 am
				5:00 am–5:30 am
Hanoi		50	12.025	3:30 am–4:00 am
				5:00 am–5:30 am
				7:00 am–7:30 am
				8:00 am–8:30 am
				3:00 pm–3:30 pm
Hanoi		50	15.004	3:00 pm–3:30 pm
				8:00 am–8:30 am
WINDWARD ISLANDS				
Grenada		10	11.975	7:00 pm–9:15 pm
YUGOSLAVIA				
Belgrade		100	6.100	1:30 pm–2:00 pm
				3:00 pm–3:30 pm
				5:00 pm–5:15 pm
Belgrade		10	7.240	1:30 pm–2:00 pm
				3:00 pm–3:30 pm
				5:00 pm–5:15 pm
Belgrade		100	9.620	10:30 am–11:00 am
				1:30 pm–2:00 pm
				3:00 pm–3:30 pm
				5:00 pm–5:15 pm
Belgrade		100	11.735	10:30 am–11:00 am
Belgrade		100	15.240	10:30 am–11:00 am
ZAIRE				
Kinshasa		50	7.115	11:00 pm–10:00 pm
Kinshasa		100	15.265	11:00 pm–10:00 pm

Location	Call Letters	Power (kW)	Freq (MHz)	Transmission Period (EST)
ZAMBIA				
Lusaka		20	6.165	Midnight–3:00 pm 9:55 pm–Midnight
Lusaka		120	7.240	Midnight–3:00 pm 9:55 pm–Midnight

Stations by Frequency

Freq (MHz)	Location	Freq (MHz)	Location
3.230	Ellice Island, Ellice	4.788	Turks & Caicos Islands,
3.230	Fiji Islands, Suva		Grand Turk
3.265	Guyana, Georgetown	4.800	Lesotho, Maseru
3.270	Dahomey, Cotonou	4.807	Reunion, St. Denis
3.300	Gabon, Libreville	4.820	Gambia, Bathurst
3.300	Honduras (British),	4.820	Honduras (Republic),
	Belize		Tegucigalpa
3.322	New Guinea,	4.845	Botswana, Gaberones
	Bougainville	4.850	Mauritius, Forest Side
3.331	Comoro Islands, Moroni	4.850	Mauritania, Nouakchott
3.355	New Caledonia,	4.870	Dahomey, Cotonou
	Noumea	4.890	Senegal, Dakar
3.356	Botswana, Gaberones	4.905	Chad, Fort Lamy
3.360	New Guinea, Milne Bay	4.907	Cambodia, Phnom Penh
3.380	Malawi, Zomba	4.915	Kenya, Nairobi
3.385	Guiana (French),	4.968	Sri Lanka (Ceylon),
	Cayenne		Colombo
3.396	Rhodesia, Salisbury	4.972	Cameroon, Yaoundé
3.910	Cape Verde Islands,	5.020	Sri Lanka (Ceylon),
	Sao Vicente		Colombo
3.980	U.S.A. (VOA)	5.035	Central African
4.715	Cape Verde Islands,		Republic, Bangui
	Sao Vicente	5.040	Burma, Rangoon
4.770	Liberia, Monrovia	5.045	Cook Islands,
4.775	Afghanistan, Kabul		Rarotonga
4.777	Gabon, Libreville	5.045	Guinea (Portuguese),
4.780	Afars & Issas, Djibouti		Bissau
4.785	Tanzania, Dar es Salaam	5.047	Togo, Lomé

Freq (MHz)	Location	Freq (MHz)	Location
5.052	Singapore, Singapore	6.075	Sri Lanka (Ceylon),
5.930	Czechoslovakia, Prague		Colombo
5.955	Germany (Dem. Rep),	6.075	Costa Rica, San José
	Berlin	6.075	Germany (Federal Rep.),
5.955	United Nations, New		Cologne
	York, N.Y.	6.080	Germany (Dem. Rep.),
5.955	U.S.A. (VOA)		Berlin
5.960	Angola, Luanda	6.080	South Africa,
5.960	Ecuador, Quito		Johannesburg
5.965	Botswana, Gaberones	6.085	Holland, Hilversum
5.965	U.S.A. (VOA)	6.090	Dominican Republic,
5.970	Canada, Montreal		Santo Domingo
5.975	Great Britain, London	6.090	Luxembourg, Villa
5.975	Paraguay, Asunción		Louvigny
5.980	El Salvador, San	6.100	Yugoslavia, Belgrade
	Salvador	6.105	Tanzania, Dar es Salaam
5.980	Romania, Bucharest	6.110	Great Britain, London
5.990	Canada, Montreal	6.120	Switzerland, Berne
5.990	Italy, Rome	6.125	Brazil, Sao Paulo
5.995	Malawi, Zomba	6.130	Ghana, Accra
5.995	U.S.A. (VOA)	6.135	Korea (Republic of),
6.005	Bolivia, La Paz		Seoul
6.005	Fiji Islands, Suva	6.135	Poland, Warsaw
6.015	U.S.A. (VOA)	6.140	Spain, Madrid
6.020	Holland, Hilversum	6.145	Germany (Federal Rep.),
6.025	Portugal, Lisbon		Cologne
6.030	Colombia, Bogotá	6.150	Romania, Bucharest
6.035	Monaco, Monte Carlo	6.150	Sudan, Omdurman
6.035	Poland, Warsaw	6.155	Austria, Vienna
6.040	U.S.A. (VOA)	6.160	U.S.A. (VOA)
6.045	Panama, David	6.165	Germany (Dem. Rep.),
6.050	Great Britain, London		Berlin
6.050	Honduras (Republic),	6.165	Zambia, Lusaka
	Tegucigalpa	6.175	Sweden, Stockholm
6.050	Mozambique, Lourenco	6.180	Great Britain, London
	Marques	6.180	Guatemala, Guatemala
6.055	Australia, Melbourne		City
6.055	Czechoslovakia, Prague	6.183	Colombia, Bogotá
6.055	Rwanda, Kigali	6.190	Romania, Bucharest
6.065	Brazil, Brasilia	6.190	Vatican State, Vatican
6.065	Sweden, Stockholm		City
6.070	Bulgaria, Sofia	6.195	Albania, Tirana

Freq (MHz)	Location	Freq (MHz)	Location
6.540	Korea (Dem. People's Rep.), Pyongyang	7.235	Germany (Federal Rep.), Cologne
7.065	Albania, Tirana	7.235	India, New Delhi
7.095	Pakistan, Karachi	7.235	Italy, Rome
7.110	Uganda, Kampala	7.240	Yugoslavia, Belgrade
7.115	Zaire, Kinshasa	7.240	Zambia, Lusaka
7.115	Thailand, Bangkok	7.245	Angola, Luanda
7.120	Burma, Rangoon	7.245	Austria, Vienna
7.120	Chad, Fort Lamy	7.245	Mauritania, Nouakchott
7.120	China (People's Rep. of), Peking	7.250	Vatican State, Vatican City
7.120	Great Britain, London	7.255	France, Paris
7.125	Guinea (Republic), Conakry	7.260	Comoro Islands, Moroni
		7.265	Pakistan, Karachi
7.130	China (Rep. of) (Taiwan), Taipei	7.265	Togo, Lomé
		7.270	Gabon, Libreville
7.130	Great Britain, London	7.270	South Africa, Johannesburg
7.135	Monaco, Monte Carlo		
7.140	Kenya, Nairobi	7.275	Italy, Rome
7.145	Ethiopia, Addis Ababa	7.275	Nigeria, Lagos
7.155	France, Paris	7.280	Tanzania, Dar es Salaam
7.155	Jordan, Amman	7.280	U.S.A. (VOA)
7.155	Niger, Niamey	7.285	Poland, Warsaw
7.160	Sarawak, Kuching	7.285	Rhodesia, Salisbury
7.165	Nepal, Katmandu	7.285	U.S.A. (VOA)
7.170	New Caledonia, Noumea	7.290	Holland, Hilversum
		7.295	Greece, Athens
7.195	Romania, Bucharest	7.300	Albania, Tirana
7.195	Uganda, Kampala	7.345	Czechoslovakia, Prague
7.195	U.S.A. (VOA)	7.470	Vietnam (Dem. Rep.), Hanoi
7.200	Sudan, Omdurman		
7.200	U.S.A. (VOA)	7.590	China (People's Rep. of), Peking
7.215	Brunei, Tutong		
7.215	Ivory Coast, Abidjan	9.009	Israel, Jerusalem
7.220	Central African Republic, Bangui	9.030	China (People's Rep. of), Peking
7.220	Hungary, Budapest	9.410	Great Britain, London
7.225	Germany (Federal Rep.), Cologne	9.460	Pakistan, Karachi
		9.475	United Arab Republic (Egypt), Cairo
7.225	Morocco, Tangier		
7.225	Romania, Bucharest	9.500	Albania, Tirana

Freq (MHz)	Location	Freq (MHz)	Location
9.500	Germany (Dem. Rep.), Berlin	9.570	Romania, Bucharest
9.505	Czechoslovakia, Prague	9.575	Italy, Rome
9.505	Dominican Republic, Santo Domingo	9.580	Australia, Melbourne
9.505	Guatemala, Guatemala City	9.585	Finland, Helsinki
		9.585	Indonesia, Jakarta
9.505	Japan, Tokyo	9.585	Sweden, Stockholm
9.505	Sudan, Omdurman	9.588	Somalia, Mogadishu
9.510	Great Britain, London	9.590	Botswana, Gaberones
9.510	Romania, Bucharest	9.590	Switzerland, Berne
9.515	Korea (Dem. People's Rep.), Pyongyang	9.595	Kuwait, Kuwait
		9.595	Nepal, Katmandu
9.515	Turkey, Ankara	9.605	Ecuador, Quito
9.515	Upper Volta, Ouagadougou	9.605	Greece, Athens
		9.610	Mauritania, Nouakchott
9.520	New Guinea, Port Moresby	9.615	Chad, Fort Lamy
		9.615	Vatican State, Vatican City
9.525	Cuba, Havana		
9.525	South Africa, Johannesburg	9.620	Germany (Federal Rep.), Cologne
9.530	Afghanistan, Kabul	9.620	Mozambique, Lourenco Marques
9.530	U.S.A. (VOA)		
9.535	Angola, Luanda	9.620	Yugoslavia, Belgrade
9.535	Switzerland, Berne	9.625	Canada, Montreal
9.540	Czechoslovakia, Prague	9.625	Israel, Jerusalem
9.540	Mongolia, Ulan Bator	9.630	Czechoslovakia, Prague
9.545	Germany (Federal Rep.), Cologne	9.630	Sweden, Stockholm
		9.635	U.S.A. (VOA)
9.545	Ghana, Accra	9.640	Korea (Republic of), Seoul
9.550	Belgium, Brussels		
9.550	Norway, Oslo	9.645	Norway, Oslo
9.555	Bolivia, La Paz	9.645	Vatican State, Vatican City
9.555	El Salvador, San Salvador		
		9.650	Guinea (Republic), Conakry
9.560	Jordan, Amman		
9.560	South Africa, Johannesburg	9.660	Malaysia, Kuala Lumpur
		9.660	U.S.S.R., Moscow
9.565	Germany (Federal Rep.), Cologne	9.668	U.S.S.R., Moscow
		9.670	U.S.A., Belmont, California
9.570	Australia, Melbourne		
9.570	India, New Delhi		

Freq (MHz)	Location	Freq (MHz)	Location
9.685	China (Rep. of) (Taiwan), Taipei	9.770	Austria, Vienna
		9.770	Haiti, Cap Haitien
6.690	Bangladesh, Dacca	9.775	Solomon Islands, Honiara
9.690	Germany (Federal Rep.), Cologne	9.780	Albania, Tirana
9.690	India, New Delhi	9.780	China (People's Rep. of), Peking
9.690	Romania, Bucharest		
9.690	U.S.A., New York, N. Y.	9.833	Hungary, Budapest
		9.850	United Arab Republic (Egypt), Cairo
9.695	Cook Islands, Rarotonga		
9.695	South Africa, Johannesburg	9.912	India, New Delhi
		10.010	Vietnam (Dem. Rep.), Hanoi
9.700	Bulgaria, Sofia		
9.700	France, Paris	10.040	Vietnam (Dem. Rep.), Hanoi
9.710	Italy, Rome		
9.710	Mauritius, Forest Side	11.620	Bangladesh, Dacca
9.715	Holland, Hilversum	11.620	India, New Delhi
9.720	Brazil, Rio de Janeiro	11.650	Bangladesh, Dacca
9.720	Sri Lanka (Ceylon), Colombo	11.650	China (People's Rep. of), Peking
9.725	Burma, Rangoon	11.675	China (People's Rep. of), Peking
9.725	Ethiopia, Addis Ababa		
9.725	Israel, Jerusalem	11.705	Lebanon, Beirut
9.730	Germany (Dem. Rep.), Berlin	11.705	New Zealand, Wellington
9.735	Germany (Federal Rep.), Cologne	11.705	Sweden, Stockholm
		11.705	Vatican State, Vatican City
9.740	Holland, Hilversum		
9.740	U.S.A. (VOA)	11.710	Australia, Melbourne
9.745	Ecuador, Quito	11.710	India, New Delhi
9.745	Iraq, Baghdad	11.715	Algeria, Algiers
9.745	Sweden, Stockholm	11.715	Switzerland, Berne
9.750	Switzerland, Berne	11.720	Greece, Athens
9.750	Tanzania, Dar es Salaam	11.720	Switzerland, Berne
9.750	U.S.A. (VOA)	11.725	Vatican State, Vatican City
9.760	Argentina, Buenos Aires		
		11.730	Holland, Hilversum
9.760	Cameroon, Yaoundé	11.735	Morocco, Tangier
9.760	Cuba, Havana	11.735	Yugoslavia, Belgrade
9.760	U.S.A. (VOA)	11.740	U.S.A. (VOA)
9.765	China (Rep. of) (Taiwan), Taipei	11.740	Vatican State, Vatican City

Freq (MHz)	Location	Freq (MHz)	Location
11.745	Ecuador, Quito	11.855	Saudi Arabia, Mecca
11.760	Cuba, Havana	11.860	Mongolia, Ulan Bator
11.760	U.S.A. (VOA)	11.865	Switzerland, Berne
11.765	Australia, Melbourne	11.870	U.S.S.R., Moscow
11.770	Mexico, Mexico City	11.875	Angola, Luanda
11.775	India, New Delhi	11.875	Belgium, Brussels
11.775	Switzerland, Berne	11.875	Nicaragua, Managua
11.780	Great Britain, London	11.880	Argentina, Buenos
11.780	Mozambique, Lourenco		Aires
	Marques	11.880	New Guinea, Port
11.780	New Zealand,		Moresby
	Wellington	11.890	Ethiopia, Addis Ababa
11.785	Germany (Federal	11.890	U.S.A., New York, N. Y.
	Rep.), Cologne	11.895	Senegal, Dakar
11.790	Afghanistan, Kabul	11.900	Malaysia, Kuala Lumpur
11.790	U.S.A. (VOA)	11.900	South Africa,
11.795	Brazil, Rio de Janeiro		Johannesburg
11.795	Germany (Federal	11.900	U.S.S.R., Moscow
	Rep.), Cologne	11.905	Germany (Federal Rep.),
11.795	Indonesia, Jakarta		Cologne
11.800	Canary Islands, Santa	11.910	Ethiopia, Addis Ababa
	Cruz	11.910	Hungary, Budapest
11.810	India, New Delhi	11.910	Thailand, Bangkok
11.810	Italy, Rome	11.915	Ecuador, Quito
11.815	Japan, Tokyo	11.920	France, Paris
11.815	Poland, Warsaw	11.920	Ivory Coast, Abidjan
11.820	Netherlands Antilles,	11.920	Philippines, Manila
	Bonaire	11.920	Romania, Bucharest
11.825	China (Rep. of)	11.920	Seychelles, Victoria
	(Taiwan), Taipei	11.920	U.S.S.R., Moscow
11.825	Colombia, Bogotá	11.925	Kuwait, Kuwait
11.830	U.S.A. (VOA)	11.925	Nigeria, Lagos
11.835	Algeria, Algiers	11.930	France, Paris
11.835	Haiti, Cap Haitien	11.930	Seychelles, Victoria
11.840	Australia, Melbourne	11.930	Sweden, Stockholm
11.840	Cuba, Havana	11.935	Portugal, Lisbon
11.840	Poland, Warsaw	11.940	Liberia, Monrovia
11.850	Ghana, Accra	11.940	Romania, Bucharest
11.850	Norway, Oslo	11.940	Singapore, Singapore
11.850	United Nations, New	11.945	Canada, Montreal
	York, N. Y.	11.945	India, New Delhi
11.850	U.S.S.R., Moscow	11.945	Paraguay, Asunción

Freq (MHz)	Location	Freq (MHz)	Location
11.950	Liberia, Monrovia	15.155	South Africa, Johannesburg
11.950	Sweden, Stockholm		
11.960	India, New Delhi	15.155	United Nations, New York, N.Y.
11.970	South Africa, Johannesburg		
		15.155	Vatican State, Vatican City
11.970	U.S.A. (VOA)		
11.975	Windward Islands, Grenada	15.165	Hungary, Budapest
		15.175	Norway, Oslo
11.990	Czechoslovakia, Prague	15.175	South Africa, Johannesburg
12.025	Vietnam (Dem. Rep.), Hanoi		
		15.185	Finland, Helsinki
12.095	Great Britain, London	15.185	Germany (Federal Rep.), Cologne
12.165	Iran, Tehran		
15.004	Vietnam (Dem. Rep.), Hanoi	15.185	Nigeria, Lagos
		15.185	Seychelles, Victoria
15.060	China (People's Rep. of), Peking	15.190	Canada, Montreal
		15.190	Congo (Republic of), Brazzaville
15.070	Great Britain, London		
15.080	India, New Delhi	15.190	India, New Delhi
15.084	Iran, Tehran	15.195	Japan, Tokyo
15.095	China (People's Rep. of), Peking	15.195	Turkey, Ankara
		15.205	India, New Delhi
15.105	Japan, Tokyo	15.205	U.S.A. (VOA)
15.110	New Zealand, Wellington	15.210	Paraguay, Asunción
		15.210	U.S.S.R., Moscow
15.115	Ecuador, Quito	15.210	Vatican State, Vatican City
15.115	Germany (Dem. Rep.), Berlin		
		15.215	U.S.A., New York, N. Y.
15.120	Sri Lanka (Ceylon), Colombo	15.220	South Africa, Johannesburg
15.120	Nigeria, Lagos	15.230	U.S.S.R., Moscow
15.120	Poland, Warsaw	15.235	India, New Delhi
15.120	Vatican State, Vatican City	15.235	Philippines, Manila
		15.240	Czechoslovakia, Prague
15.125	China (Rep.of) (Taiwan), Taipei	15.240	Sweden, Stockholm
		15.240	Yugoslavia, Belgrade
15.145	Austria, Vienna	15.250	Romania, Bucharest
15.150	Korea (Dem. People's Rep.), Pyongyang	15.255	Netherlands Antilles, Bonaire
		15.260	Great Britain, London
15.155	Cuba, Havana	15.265	Afghanistan, Kabul
15.155	Korea (Republic of), Seoul	15.265	Zaire, Kinshasa

Freq (MHz)	Location	Freq (MHz)	Location
15.270	Cuba, Havana	15.510	China (People's Rep.
15.270	Seychelles, Victoria		of), Peking
15.275	Poland, Warsaw	15.520	Bangladesh, Dacca
15.280	Malaysia, Kuala Lumpur	17.655	United Arab Republic
15.280	U.S.A., Belmont,		(Egypt), Cairo
	California	17.710	Romania, Bucharest
15.285	Cuba, Havana	17.715	Australia, Melbourne
15.285	Ghana, Accra	17.715	China (People's Rep.
15.295	France, Paris		of), Peking
15.295	U.S.S.R., Moscow	17.720	China (Rep. of)
15.300	Ecuador, Quito		(Taiwan), Taipei
15.300	Philippines, Manila	17.720	France, Paris
15.305	Switzerland, Berne	17.725	Mali, Bamako
15.310	Guinea (Republic),	17.725	United Arab Republic
	Conakry		(Egypt), Cairo
15.315	Ethiopia, Addis Ababa	17.730	France, Paris
15.320	Australia, Melbourne	17.730	Malagasy (Republic),
15.320	Germany (Federal		Tananarive
	Rep.), Cologne	17.730	U.S.S.R., Moscow
15.325	Canada, Montreal	17.735	China (People's Rep.
15.330	Italy, Rome		of), Peking
15.335	India, New Delhi	17.750	Kuwait, Kuwait
15.345	China (Rep. of)	17.770	New Zealand,
	(Taiwan), Taipei		Wellington
15.345	Kuwait, Kuwait	17.775	India, New Delhi
15.365	Canary Islands, Santa	17.775	U.S.S.R., Moscow
	Cruz	17.780	Afghanistan, Kabul
15.370	China (Rep. of)	17.780	China (Rep. of)
	(Taiwan), Taipei		(Taiwan), Taipei
15.385	China (People's Rep.	17.780	Mongolia, Ulan Bator
	of), Peking	17.785	Japan, Tokyo
15.390	Germany (Dem. Rep.),	17.785	U.S.A. (VOA)
	Berlin	17.790	Great Britain, London
15.390	Japan, Tokyo	17.795	Australia, Melbourne
15.420	Algeria, Algiers	17.795	Italy, Rome
15.430	Korea (Republic of),	17.795	Switzerland, Berne
	Seoul	17.850	South Africa,
15.430	Switzerland, Berne		Johannesburg
15.435	Tanzania, Dar es Salaam	17.810	Philippines, Manila
15.440	Philippines, Manila	17.815	South Africa,
15.445	Mongolia, Ulan Bator		Johannesburg
		17.820	Canada, Montreal

Freq (MHz)	Location	Freq (MHz)	Location
17.820	South Africa, Johannesburg	21.485	Germany (Dem. Rep.), Berlin
17.820	Turkey, Ankara	21.485	Australia, Melbourne
17.830	Sri Lanka (Ceylon), Colombo	21.485	U.S.A. (VOA)
17.830	Holland, Hilversum	21.485	Vatican State, Vatican City
17.830	Switzerland, Berne	21.495	Portugal, Lisbon
17.830	United Nations, New York, N. Y.	21.500	Congo (Republic of), Brazzaville
17.835	Mexico, Mexico City	21.505	Sweden, Stockholm
17.840	Czecholslovakia, Prague	21.515	Philippines, Manila
17.840	Hungary, Budapest	21.520	South Africa, Johannesburg
17.840	Sweden, Stockholm	21.520	Switzerland, Berne
17.840	Vatican State, Vatican City	21.525	U.S.A., New York, N.Y.
17.845	Switzerland, Berne	21.535	South Africa, Johannesburg
17.845	U.S.A., New York, N. Y.	21.540	Germany (Dem. Rep.), Berlin
17.850	Romania, Bucharest		
17.855	Japan, Tokyo	21.540	Germany (Federal Rep.), Cologne
17.865	U.S.S.R., Moscow		
17.870	Ghana, Accra	21.545	Ghana, Accra
17.875	Germany (Federal Rep.), Cologne	21.545	South Africa, Johannesburg
17.880	Ecuador, Quito	21.555	India, New Delhi
17.880	Germany (Dem. Rep.), Berlin	21.580	France, Paris
17.880	Japan, Tokyo	21.585	Switzerland, Berne
17.880	Portugal, Lisbon	21.590	Pakistan, Karachi
17.890	China (Rep. of) (Taiwan), Taipei	21.600	Germany (Dem. Rep.), Berlin
17.895	Portugal, Lisbon	21.605	Switzerland, Berne
17.920	United Arab Republic (Egypt), Cairo	21.650	U.S.A. (VOA)
		21.655	Norway, Oslo
17.935	Bangladesh, Dacca	21.685	Hungary, Budapest
21.460	Ecuador, Quito	21.690	Sweden, Stockholm
21.465	Germany (Dem. Rep.), Berlin	21.700	Czechoslovakia, Prague
		21.705	Mexico, Mexico City
21.470	Great Britain, London	21.740	Australia, Melbourne
21.480	South Africa, Johannesburg	25.790	South Africa, Johannesburg

Stations by Time Period

Part 1—Midnight—4:00 am EST

Location	Freq (MHz)	Actual Time
AFARS & ISSAS		
Djibouti	4.780	Midnight—1:00 am
ALGERIA		
Algiers	11.715	1:00 am—4:00 am
Algiers	11.835	1:00 am—4:00 am
Algiers	15.420	1:00 am—4:00 am
ANGOLA		
Luanda	5.960	Midnight—3:00 am
Luanda	7.245	Midnight—4:00 am
Luanda	9.535	Midnight—4:00 am
Luanda	11.875	3:00 am—4:00 am
ARGENTINA		
Buenos Aires	9.760	Midnight—4:00 am
AUSTRALIA		
Melbourne	9.570	1:45 am—2:45 am
Melbourne	11.765	1:45 am—2:45 am
Melbourne	17.715	Midnight—2:00 am
Melbourne	21.485	1:15 am—3:00 am
AUSTRIA		
Vienna	6.155	Midnight—4:00 am
Vienna	7.245	1:00 am—3:00 am

Location	Freq (MHz)	Actual Time
BRAZIL		
Brasilia	6.065	3:00 am–4:00 am
Rio de Janeiro	9.720	3:00 am–4:00 am
Sao Paulo	6.125	3:00 am–4:00 am
BURMA		
Rangoon	9.725	2:00 am–2:30 am
CAMBODIA		
Phnom Penh	4.907	12:45 am–1:00 am
CAMEROON		
Yaoundé	4.972	12:30 am–1:00 am
CANADA		
Montreal	5.970	3:30 am–4:00 am
Montreal	5.990	2:15 am–2:45 am
Montreal	9.625	2:15 am–2:45 am
		3:30 am–4:00 am
Montreal	15.325	2:15 am–2:45 am
Montreal	17.820	2:15 am–2:45 am
CAPE VERDE ISLANDS		
Sao Vicente	4.715	2:00 am–4:00 am
CENTRAL AFRICAN REPUBLIC		
Bangui	7.220	2:30 am–4:00 am
CHAD		
Fort Lamy	4.905	Midnight–1:30 am
CONGO (Republic of)		
Brazzaville	15.190	12:15 am–12:30 am
Brazzaville	21.500	12:15 am–12:30 am
COOK ISLANDS		
Rarotonga	5.045	Midnight–3:00 am
COSTA RICA		
San José	6.075	Midnight–4:00 am

Location	Freq (MHz)	Actual Time
CUBA		
Havana	9.525	1:30 am–3:00 am
Havana	11.760	Midnight–1:00 am
Havana	11.840	Midnight–1:00 am
CZECHOSLOVAKIA		
Prague	6.055	2:00 am–3:00 am
Prague	9.505	2:00 am–3:00 am
Prague	21.700	2:00 am–3:00 am
DAHOMEY		
Cotonou	3.270	12:15 am–12:30 am
		1:15 am–1:30 am
Cotonou	4.870	12:15 am–12:30 am
		1:15 am–1:30 am
ECUADOR		
Quito	5.960	2:15 am–4:00 am
Quito	9.745	2:15 am–4:00 am
Quito	11.915	2:15 am–4:00 am
Quito	15.115	Midnight–4:00 am
ELLICE ISLAND		
Ellice	3.230	2:00 am–3:00 am
ETHIOPIA		
Addis Ababa	11.890	12:30 am–1:00 am
FIJI ISLANDS		
Suva	3.230	Midnight–4:00 am
FRANCE		
Paris	7.155	12:15 am–12:30 am
Paris	7.255	12:15 am–12:30 am
Paris	9.700	12:15 am–12:30 am
Paris	11.920	12:15 am–12:30 am
Paris	11.930	12:15 am–12:30 am
Paris	15.295	12:15 am–12:30 am
Paris	17.730	12:15 am–12:30 am
GABON		
Libreville	3.300	Midnight–1:30 am

Location	Freq (MHz)	Actual Time
GABON (cont)		
Libreville	4.777	Midnight–1:30 am
Libreville	7.270	1:30 am–4:00 am
GAMBIA		
Bathurst	4.820	1:30 am–3:00 am
GERMANY (Democratic Rep.)		
Berlin	9.500	1:15 am–2:00 am
Berlin	21.465	1:45 am–2:30 am
GERMANY (Federal Rep.)		
Cologne	6.075	Midnight–1:00 am
Cologne	6.145	Midnight–1:00 am
Cologne	9.545	Midnight–1:00 am
Cologne	9.620	1:00 am–1:30 am
Cologne	9.690	12:45 am–1:30 am
Cologne	11.785	1:00 am–1:30 am
Cologne	11.905	12:45 am–1:30 am
Cologne	15.320	1:00 am–1:30 am
GREAT BRITAIN		
London	6.050	Midnight–2:30 am
London	6.110	Midnight–2:00 am
London	6.180	Midnight–2:30 am
London	9.410	Midnight–3:00 am
London	17.790	1:00 am–2:30 am
GUINEA (Portuguese)		
Bissau	5.045	1:00 am–4:00 am
GUINEA (Republic)		
Conakry	7.125	1:00 am–3:30 am
Conakry	15.310	1:00 am–3:30 am
HOLLAND		
Hilversum	9.715	Midnight–1:30 am
		3:00 am–4:00 am
Hilversum	11.730	Midnight–3:00 am
HUNGARY		
Budapest	9.833	Midnight–12:30 am

Location	Freq (MHz)	Actual Time
HUNGARY (cont)		
Budapest	11.910	Midnight–12:30 am
Budapest	15.165	Midnight–12:30 am
IVORY COAST		
Abidjan	7.215	1:00 am–3:00 am
Abidjan	11.920	1:00 am–3:00 am
JAPAN		
Tokyo	9.505	2:00 am–2:15 am
		3:00 am–3:15 am
Tokyo	15.195	2:00 am–2:15 am
		3:00 am–3:15 am
Tokyo	17.855	Midnight–12:15 am
		1:00 am–1:15 am
		2:00 am–2:15 am
		3:00 am–3:15 am
Tokyo	17.880	Midnight–12:15 am
		1:00 am–1:15 am
KENYA		
Nairobi	4.915	Midnight–1:30 am
KOREA (Dem. People's Rep.)		
Pyongyang	6.540	3:00 am–4:00 am
Pyongyang	15.150	3:00 am–4:00 am
KOREA (Republic of)		
Seoul	6.135	3:00 am–3:30 am
Seoul	9.640	Midnight–12:30 am
Seoul	15.155	1:30 am–2:00 am
KUWAIT		
Kuwait	17.750	Midnight–1:00 am
LESOTHO		
Maseru	4.800	Midnight–2:00 am
LIBERIA		
Monrovia	4.770	1:15 am–3:15 am
Monrovia	11.950	1:30 am–2:15 am

Location	Freq (MHz)	Actual Time
MALAWI		
Zomba	3.380	Midnight–12:30 am
		3:00 am–4:00 am
Zomba	5.995	Midnight–12:30 am
		3:00 am–4:00 am
MALAYSIA		
Kuala Lumpur	9.660	12:30 am–1:30 am
Kuala Lumpur	11.900	12:30 am–4:00 am
Kuala Lumpur	15.280	12:30 am–4:00 am
MAURITANIA		
Nouakchott	4.850	2:00 am–3:00 am
MAURITIUS		
Forest Side	9.710	Midnight–4:00 am
MEXICO		
Mexico City	11.770	Midnight–1:00 am
MONACO		
Monte Carlo	6.035	Midnight–4:00 am
Monte Carlo	7.135	Midnight–4:00 am
MOZAMBIQUE		
Lourenco Marques	6.050	Midnight–4:00 am
Lourenco Marques	9.620	2:00 am–4:00 am
Lourenco Marques	11.780	Midnight–4:00 am
NEPAL		
Katmandu	7.165	2:20 am–4:00 am
Katmandu	9.595	2:00 am–4:00 am
NEW CALEDONIA		
Noumea	3.355	1:00 am–4:00 am
Noumea	7.170	1:00 am–4:00 am
NEW GUINEA		
Bougainville	3.322	2:00 am–4:00 am
Milne Bay	3.360	2:30 am–4:00 am
Port Moresby	9.520	Midnight–2:00 am
Port Morseby	11.880	Midnight–12:30 am

Location	Freq (MHz)	Actual Time
NEW ZEALAND		
Wellington	11.780	1:00 am–4:00 am
NIGERIA		
Lagos	7.275	12:30 am–2:30 am
Lagos	15.120	12:30 am–2:30 am
Lagos	15.185	12:30 am–2:30 am
NORWAY		
Oslo	9.645	Midnight–12:30 am
Oslo	11.850	2:00 am–2:30 am
Oslo	15.175	2:00 am–2:30 am
Oslo	21.655	2:00 am–2:30 am
PAKISTAN		
Karachi	21.590	Midnight–4:00 am
PHILIPPINES		
Manila	15.440	Midnight–1:30 am
Manila	17.810	Midnight–1:00 am
Manila	21.515	Midnight–1:30 am
PORTUGAL		
Lisbon	17.880	2:30 am–4:00 am
Lisbon	21.495	2:30 am–4:00 am
REUNION		
St. Denis	4.807	Midnight–4:00 am
RHODESIA		
Salisbury	3.396	Midnight–1:00 am
Salisbury	7.285	1:00 am–4:00 am
ROMANIA		
Bucharest	15.250	1:45 am–2:15 am
Bucharest	17.850	1:45 am–2:15 am
RWANDA		
Kigali	6.055	Midnight–1:00 am
SAUDI ARABIA		
Mecca	11.855	Midnight–1:00 am

Location	Freq (MHz)	Actual Time
SINGAPORE		
Singapore	5.052	Midnight–4:00 am
Singapore	11.940	Midnight–4:00 am
SOLOMAN ISLANDS		
Honiara	9.775	2:00 am–4:00 am
SOUTH AFRICA		
Johannesburg	7.270	Midnight–2:00 am
Johannesburg	9.525	Midnight–12:15 am
Johannesburg	11.900	Midnight–12:15 am
		1:30 am–1:45 am
Johannesburg	15.220	Midnight–1:45 am
Johannesburg	17.805	Midnight–12:15 am
		1:30 am–1:45 am
Johannesburg	17.815	3:00 am–4:00 am
Johannesburg	21.535	1:30 am–1:45 am
Johannesburg	21.545	3:00 am–4:00 am
SWEDEN		
Stockholm	9.585	12:15 am–12:45 am
Stockholm	17.840	12:15 am–12:45 am
SWITZERLAND		
Berne	9.590	2:00 am–2:30 am
Berne	11.775	2:00 am–2:30 am
Berne	11.865	3:45 am–4:00 am
Berne	15.305	3:45 am–4:00 am
THAILAND		
Bangkok	7.115	Midnight–12:15 am
Bangkok	11.910	Midnight–12:15 am
UNITED NATIONS		
New York	5.955	3:45 am–4:00 am
New York	11.850	3:45 am–4:00 am
UNITED STATES OF AMERICA		
Washington, D. C. (VOA)	3.980	Midnight–2:30 am
Washington, D. C. (VOA)	5.955	Midnight–2:30 am
Washington, D. C. (VOA)	5.965	Midnight–1:00 am
Washington, D. C. (VOA)	5.995	Midnight–2:30 am

Location	Freq (MHz)	Actual Time

UNITED STATES OF AMERICA (cont)

Washington, D. C. (VOA)	6.160	Midnight–2:30 am
Washington, D. C. (VOA)	7.195	Midnight–2:30 am
Washington, D. C. (VOA)	7.200	Midnight–2:30 am
Washington, D. C. (VOA)	7.280	Midnight–2:30 am
Washington, D. C. (VOA)	9.530	Midnight–2:30 am
Washington, D. C. (VOA)	9.635	Midnight–2:30 am
Washington, D. C. (VOA)	9.740	Midnight–2:30 am
Washington, D. C. (VOA)	9.750	Midnight–2:30 am
Washington, D. C. (VOA)	11.740	1:00 am–2:30 am
Washington, D. C. (VOA)	11.760	Midnight–2:30 am
Washington, D. C. (VOA)	11.830	1:00 am–2:00 am
Washington, D. C. (VOA)	11.970	1:00 am–2:00 am

UPPER VOLTA

Ouagadougou	9.515	1:00 am–3:00 am

U.S.S.R.

Moscow	9.660	Midnight–12:30 am
Moscow	9.668	Midnight–12:30 am
Moscow	11.850	Midnight–12:30 am
Moscow	11.870	Midnight–12:30 am
Moscow	11.900	Midnight–12:30 am
Moscow	11.920	Midnight–12:30 am
Moscow	15.210	Midnight–12:30 am
Moscow	17.730	Midnight–12:30 am
Moscow	17.865	Midnight–1:00 am

VATICAN STATE

Vatican City	7.250	2:30 am–3:00 am
Vatican City	11.740	2:30 am–3:30 am
Vatican City	15.120	2:30 am–3:30 am

VIETNAM (Democratic Rep.)

Hanoi	10.040	Midnight–12:30 am
		3:30 am–4:00 am
Hanoi	12.025	3:30 am–4:00 am

ZAIRE

Kinshasa	7.115	Midnight–4:00 am
Kinshasa	15.265	Midnight–4:00 am

Location	Freq (MHz)	Actual Time

ZAMBIA
Lusaka   6.165   Midnight—4:00 am
Lusaka   7.240   Midnight—4:00 am

Part 2—4:00 am—8:00 am EST

ALBANIA
Tirana   9.500   6:00 am—6:30 am

ALGERIA
Algiers   11.715   4:00 am—8:00 am
Algiers   11.835   4:00 am—8:00 am
Algiers   15.420   4:00 am—8:00 am

ANGOLA
Luanda   5.960   6:00 am—8:00 am
Luanda   7.245   4:00 am—8:00 am
Luanda   9.535   4:00 am—8:00 am
Luanda   11.875   4:00 am—8:00 am

ARGENTINA
Buenos Aires   9.760   4:00 am—5:00 am
Buenos Aires   11.880   7:00 am—8:00 am

AUSTRALIA
Melbourne   9.580   7:15 am—8:00 am
Melbourne   11.710   7:15 am—8:00 am

AUSTRIA
Vienna   6.155   4:00 am—7:00 am

BANGLADESH
Dacca   11.620   7:30 am—8:00 am
Dacca   17.935   7:30 am—8:00 am

BOLIVIA
La Paz   6.005   4:00 am—8:00 am
La Paz   9.555   6:00 am—8:00 am

Location	Freq (MHz)	Actual Time
BOTSWANA		
Gaberones	5.965	6:30 am–6:45 am
Gaberones	9.590	6:30 am–6:45 am
BRAZIL		
Brasilia	6.065	4:00 am–8:00 am
Rio de Janeiro	9.720	4:00 am–8:00 am
Rio de Janeiro	11.795	4:00 am–8:00 am
Sao Paulo	6.125	4:00 am–8:00 am
BRUNEI		
Tutong	7.215	7:00 am–8:00 am
CAMBODIA		
Phnom Penh	4.907	7:45 am–8:00 am
CANADA		
Montreal	5.970	4:00 am–4:30 am
Montreal	9.625	4:00 am–4:30 am
		7:15 am–8:00 am
CAPE VERDE ISLANDS		
Sao Vicente	4.715	6:00 am–8:00 am
CENTRAL AFRICAN REPUBLIC		
Bangui	7.220	4:00 am–8:00 am
CHAD		
Fort Lamy	7.120	7:30 am–8:00 am
Fort Lamy	9.615	7:30 am–8:00 am
CHINA (People's Rep. of)		
Peking	15.510	7:00 am–8:00 am
COLOMBIA		
Bogotá	6.030	6:00 am–8:00 am
Bogotá	6.183	6:00 am–8:00 am
Bogotá	11.825	6:00 am–8:00 am
COMORO ISLANDS		
Moroni	7.260	4:00 am–6:00 am

Location	Freq (MHz)	Actual Time
CONGO (Republic of)		
Brazzaville	15.190	6:00 am–6:30 am
Brazzaville	21.500	6:00 am–6:30 am
COSTA RICA		
San José	6.075	4:00 am–8:00 am
DAHOMEY		
Cotonou	3.270	7:00 am–7:15 am
Cotonou	4.870	7:00 am–7:15 am
DOMINICAN REPUBLIC		
Santo Domingo	6.090	6:00 am–8:00 am
Santo Domingo	9.505	6:00 am–8:00 am
ECUADOR		
Quito	5.960	4:00 am–5:00 am
Quito	9.745	4:00 am–6:00 am
Quito	11.915	4:00 am–6:00 am
Quito	15.115	4:00 am–5:00 am
		7:45 am–8:00 am
Quito	17.880	7:45 am–8:00 am
Quito	21.460	7:45 am–8:00 am
FIJI ISLANDS		
Suva	3.230	4:00 am–4:15 am
FRANCE		
Paris	15.295	6:00 am–6:15 am
Paris	17.720	6:00 am–6:15 am
Paris	21.580	6:00 am–6:15 am
GABON		
Libreville	7.270	4:00 am–8:00 am
GAMBIA		
Bathurst	4.820	7:00 am–8:00 am
GERMANY (Democratic Rep.)		
Berlin	15.115	7:00 am–7:45 am
Berlin	21.540	7:00 am–7:45 am

Location	Freq (MHz)	Actual Time
GERMANY (Federal Rep.)		
Cologne	11.795	4:20 am–5:20 am
Cologne	15.185	4:20 am–5:20 am
Cologne	17.875	6:15 am–6:45 am
Cologne	21.540	6:15 am–6:45 am
GREAT BRITAIN		
London	12.095	4:00 am–8:00 am
London	15.070	4:00 am–8:00 am
London	17.790	7:00 am–8:00 am
London	21.470	4:00 am–8:00 am
GUATEMALA		
Guatemala City	6.180	7:00 am–8:00 am
Guatemala City	9.505	6:00 am–8:00 am
GUINEA (Portuguese)		
Bissau	5.045	4:00 am–8:00 am
GUYANA		
Georgetown	3.265	4:15 am–8:00 am
HAITI		
Cap Haitien	9.770	6:00 am–8:00 am
Cap Haitien	11.835	6:00 am–8:00 am
HOLLAND		
Hilversum	6.020	4:30 am–6:00 am
Hilversum	9.715	4:00 am–4:30 am
HONDURAS (Republic)		
Tegucigalpa	4.820	5:00 am–8:00 am
INDIA		
New Delhi	11.775	5:00 am–6:00 am
New Delhi	15.190	5:00 am–6:00 am
New Delhi	15.205	5:00 am–6:00 am
New Delhi	17.775	5:00 am–6:00 am
New Delhi	21.555	5:00 am–6:00 am
INDONESIA		
Jakarta	11.795	4:00 am–4:30 am

Location	Freq (MHz)	Actual Time
JAPAN		
Tokyo	9.505	4:00 am—4:15 am
		5:00 am—5:15 am
		7:00 am—7:15 am
Tokyo	11.815	5:00 am—5:15 am
		7:00 am—7:15 am
Tokyo	15.195	4:00 am—4:15 am
Tokyo	15.390	5:00 am—5:15 am
		7:00 am—7:15 am
Tokyo	17.855	4:00 am—4:15 am
JORDAN		
Amman	7.155	5:00 am—8:00 am
KENYA		
Nairobi	4.915	4:00 am—6:00 am
Nairobi	7.140	4:00 am—6:00 am
KOREA (Dem. People's Rep.)		
Pyongyang	15.150	6:00 am—7:00 am
KOREA (Republic of)		
Seoul	6.135	6:00 am—6:30 am
Seoul	9.640	4:00 am—4:30 am
		6:00 am—6:30 am
Seoul	15.430	4:00 am—4:30 am
LESOTHO		
Maseru	4.800	5:30 am—7:00 am
MALAYSIA		
Kuala Lumpur	9.660	4:30 am—8:00 am
Kuala Lumpur	11.900	4:30 am—8:00 am
Kuala Lumpur	15.280	4:30 am—8:00 am
MAURITANIA		
Nouakchott	7.245	7:00 am—8:00 am
Nouakchott	9.610	7:00 am—8:00 am
MAURITIUS		
Forest Side	9.710	4:00 am—8:00 am

Location	Freq (MHz)	Actual Time
MONACO		
Monte Carlo	6.035	4:00 am—8:00 am
Monte Carlo	7.135	4:00 am—8:00 am
MONGOLIA		
Ulan Bator	15.445	7:20 am—7:40 am
Ulan Bator	17.780	7:20 am—7:40 am
MOZAMBIQUE		
Lourenco Marques	6.050	4:00 am—8:00 am
Lourenco Marques	9.620	4:00 am—8:00 am
Lourenco Marques	11.780	4:00 am—8:00 am
NEPAL		
Katmandu	7.165	4:00 am—4:30 am
NETHERLANDS ANTILLES		
Bonaire	11.820	6:00 am—7:30 am
Bonaire	15.255	7:30 am—8:00 am
NEW CALEDONIA		
Noumea	3.355	4:00 am—6:00 am
Noumea	7.170	4:00 am—6:00 am
NEW GUINEA		
Bougainville	3.322	4:00 am—7:00 am
Milne Bay	3.360	4:00 am—7:00 am
NEW ZEALAND		
Wellington	11.705	4:00 am—7:00 am
NIGER		
Niamey	7.155	6:30 am—8:00 am
NORWAY		
Oslo	21.655	6:00 am—6:30 am
PAKISTAN		
Karachi	21.590	4:00 am—4:15 am
PANAMA		
David	6.045	6:60 am—8:00 am

Location	Freq (MHz)	Actual Time
PARAGUAY		
Asunción	5.975	7:00 am–8:00 am
Asunción	11.945	4:30 am–8:00 am
Asunción	15.210	4:30 am–8:00 am
PHILIPPINES		
Manila	11.920	6:45 am–8:00 am
REUNION		
St. Denis	4.807	4:00 am–8:00 am
RHODESIA		
Salisbury	7.285	4:00 am–8:00 am
ROMANIA		
Bucharest	11.920	6:00 am–6:30 am
Bucharest	15.250	6:00 am–6:30 am
Bucharest	17.850	6:00 am–6:30 am
RWANDA		
Kigali	6.055	4:00 am–7:00 am
SARAWAK		
Kuching	7.160	6:00 am–6:30 am
SAUDI ARABIA		
Mecca	11.855	6:00 am–7:00 am
SINGAPORE		
Singapore	5.052	4:00 am–8:00 am
Singapore	11.940	4:00 am–8:00 am
SOLOMON ISLANDS		
Honiara	9.775	4:00 am–6:15 am
SOUTH AFRICA		
Johannesburg	11.900	6:00 am–8:00 am
Johannesburg	15.220	5:00 am–8:00 am
Johannesburg	17.820	5:00 am–6:00 am
Johannesburg	21.520	5:00 am–6:00 am
Johannesburg	21.535	4:30 am–8:00 am
Johannesburg	25.790	4:30 am–4:45 am

Location	Freq (MHz)	Actual Time
SRI LANKA (Ceylon)		
Colombo	4.968	7:30 am–8:00 am
Colombo	5.020	6:30 am–8:00 am
Colombo	17.830	6:00 am–7:00 am
SWEDEN		
Stockholm	6.065	7:30 am–8:00 am
Stockholm	9.630	6:00 am–6:30 am
		7:30 am–8:00 am
Stockholm	21.690	6:00 am–6:30 am
SWITZERLAND		
Berne	11.865	4:00 am–4:15 am
Berne	15.305	4:00 am–4:15 am
Berne	15.430	6:00 am–6:30 am
Berne	17.795	6:00 am–6:30 am
Berne	21.520	6:00 am–6:30 am
Berne	21.585	6:00 am–6:30 am
TANZANIA		
Dar es Salaam	7.280	4:00 am–5:30 am
Dar es Salaam	9.750	4:00 am–5:30 am
THAILAND		
Bangkok	7.115	5:30 am–6:30 am
Bangkok	11.910	5:30 am–6:30 am
TOGO		
Lomé	7.265	7:45 am–8:00 am
UGANDA		
Kampala	7.110	4:00 am–4:15 am
Kampala	7.195	4:00 am–4:15 am
UPPER VOLTA		
Ouagadougou	9.515	7:00 am–8:00 am
VATICAN STATE		
Vatican City	9.615	7:50 am–8:00 am
Vatican City	11.725	7:50 am–8:00 am
Vatican City	15.155	7:50 am–8:00 am

Location	Freq (MHz)	Actual Time
VATICAN STATE (cont)		
Vatican City	17.840	5:00 am–7:00 am
Vatican City	21.485	5:00 am–7:00 am
VIETNAM (Democratic Rep.)		
Hanoi	7.470	6:30 am–7:00 am
Hanoi	10.040	5:00 am–5:30 am
Hanoi	12.025	5:00 am–5:30 am
		7:00 am–7:30 am
ZAIRE		
Kinshasa	7.115	4:00 am–8:00 am
Kinshasa	15.265	4:00 am–8:00 am
ZAMBIA		
Lusaka	6.165	4:00 am–8:00 am
Lusaka	7.240	4:00 am–8:00 am

Part 3—8:00 am—Noon EST

Location	Freq (MHz)	Actual Time
AFGHANISTAN		
Kabul	4.775	9:00 am–9:30 am
ALGERIA		
Algiers	11.715	8:00 am–Noon
Algiers	11.835	8:00 am–Noon
Algiers	15.420	8:00 am–Noon
ANGOLA		
Luanda	7.245	8:00 am–Noon
Luanda	9.535	8:00 am–Noon
Luanda	11.875	8:00 am–Noon
ARGENTINA		
Buenos Aires	11.880	8:00 am–Noon
AUSTRALIA		
Melbourne	9.580	8:00 am–8:15 am
Melbourne	11.710	8:00 am–8:15 am

Location	Freq (MHz)	Actual Time
BANGLADESH		
Dacca	9.690	9:30 am–10:00 am
Dacca	15.520	9:30 am–10:00 am
BOLIVIA		
La Paz	6.005	8:00 am–Noon
La Paz	9.555	10:00 am–Noon
BOTSWANA		
Gaberones	3.356	11:00 am–11:30 am
Gaberones	4.845	11:00 am–11:30 am
BRAZIL		
Brasilia	6.065	8:00 am–Noon
Rio de Janeiro	9.720	8:00 am–Noon
Rio de Janeiro	11.795	8:00 am–Noon
Sao Paulo	6.125	8:00 am–Noon
BRUNEI		
Tutong	7.215	8:00 am–9:30 am
BURMA		
Rangoon	5.040	9:30 am–11:00 am
CAMEROON		
Yaoundé	4.972	8:30 am–9:45 am
Yaoundé	9.760	8:00 am–9:30 am
CANADA		
Montreal	9.625	8:00 am–8:15 am
CAPE VERDE ISLANDS		
Sao Vicente	3.910	10:00 am–11:00 am
CENTRAL AFRICAN REPUBLIC		
Bangui	5.035	11:30 am–Noon
Bangui	7.220	8:00 am–11:30 am
CHAD		
Fort Lamy	4.905	11:15 am–Noon
Fort Lamy	7.120	8:00 am–11:15 am
Fort Lamy	9.615	8:00 am–11:15 am

Location	Freq (MHz)	Actual Time
CHINA (People's Rep of)		
Peking	15.510	8:00 am–9:00 am
COLOMBIA		
Bogotá	6.030	8:00 am–Noon
Bogotá	6.183	8:00 am–Noon
Bogotá	11.825	8:00 am–Noon
COMORO ISLANDS		
Moroni	3.331	10:30 am–Noon
COOK ISLANDS		
Rarotonga	5.045	11:30 am–Noon
COSTA RICA		
San José	6.075	8:00 am–Noon
CZECHOSLOVAKIA		
Prague	5.930	11:30 am–Noon
Prague	6.055	10:00 am–11:30 am
Prague	7.345	11:30 am–Noon
Prague	9.505	10:00 am–10:30 am
Prague	15.240	10:30 am–11:30 am
Prague	17.840	10:30 am–11:30 am
DAHOMEY		
Cotonou	3.270	11:45 am–Noon
Cotonou	4.870	11:45 am–Noon
DOMINICAN REPUBLIC		
Santo Domingo	6.090	8:00 am–Noon
Santo Domingo	9.505	8:00 am–Noon
ECUADOR		
Quito	15.115	8:00 am–11:30 am
Quito	17.880	8:00 am–11:30 am
Quito	21.460	8:00 am–11:30 am
ETHIOPIA		
Addis Ababa	15.315	8:00 am–9:00 am

Location	Freq (MHz)	Actual Time
FINLAND		
Helsinki	15.185	11:00 am–Noon
GABON		
Libreville	4.777	11:30 am–Noon
Libreville	7.270	8:00 am–11:30 am
GAMBIA		
Bathurst	4.820	8:00 am–Noon
GERMANY (Democratic Rep.)		
Berlin	17.880	10:30 am–11:15 am
Berlin	21.540	9:00 am–9:45 am
Berlin	21.600	8:15 am–9:00 am
GHANA		
Accra	6.130	9:00 am–Noon
Accra	17.870	9:00 am–10:00 am
Accra	21.545	9:45 am–10:30 am
GREAT BRITAIN		
London	9.410	10:00 am–Noon
London	12.095	8:00 am–Noon
London	15.070	8:00 am–Noon
London	15.260	10:00 am–10:15 am
London	17.790	8:00 am–Noon
London	21.470	8:00 am–Noon
GUATEMALA		
Guatemala City	6.180	8:00 am–Noon
Guatemala City	9.505	8:00 am–Noon
GUINEA (Portuguese)		
Bissau	5.045	8:00 am–Noon
GUINEA (Republic)		
Conakry	7.125	11:00 am–Noon
Conakry	15.310	11:00 am–Noon
GUIANA (French)		
Cayenne	3.385	10:00 am–Noon

Location	Freq (MHz)	Actual Time
GUYANA		
Georgetown	3.265	8:00 am–Noon
HAITI		
Cap Haitien	9.770	8:00 am–Noon
Cap Haitien	11.835	8:00 am–Noon
HOLLAND		
Hilversum	6.020	9:00 am–10:30 am
HONDURAS (Republic)		
Tegucigalpa	4.820	8:00 am–Noon
Tegucigalpa	6.050	8:00 am–Noon
INDIA		
New Delhi	11.810	8:00 am–10:00 am
New Delhi	15.335	8:00 am–10:00 am
INDONESIA		
Jakarta	9.585	9:30 am–10:30 am
Jakarta	11.795	9:30 am–10:30 am
JAPAN		
Tokyo	9.505	8:00 am–8:15 am
		9:00 am–9:15 am
		11:00 am–11:15 am
Tokyo	11.815	8:00 am–8:15 am
		9:00 am–9:15 am
		11:00 am–11:15 am
Tokyo	15.390	8:00 am–8:15 am
		9:00 am–9:15 am
		11:00 am–11:15 am
JORDAN		
Amman	7.155	8:00 am–8:30 am
Amman	9.560	9:00 am–Noon
KENYA		
Nairobi	4.915	8:00 am–Noon
KOREA (Dem. People's Rep.)		
Pyongyang	9.515	8:00 am–10:00 am

Location	Freq (MHz)	Actual Time
KOREA (Dem. People's Rep.) (cont)		
Pyongyang	15.150	8:00 am—10:00 am
KOREA (Republic of)		
Seoul	15.430	8:30 am—9:00 am
KUWAIT		
Kuwait	9.595	11:00 am—Noon
Kuwait	15.345	11:00 am—Noon
LESOTHO		
Maseru	4.800	10:00 am—Noon
LIBERIA		
Monrovia	4.770	11:15 am—Noon
MALAGASY (Republic)		
Tananarive	17.730	8:30 am—9:30 am
MALAYSIA		
Kuala Lumpur	9.660	8:00 am—11:30 am
Kuala Lumpur	11.900	8:00 am—11:30 am
Kuala Lumpur	15.280	8:00 am—11:30 am
MALI		
Bamako	17.725	10:00 am—Noon
MAURITANIA		
Nouakchott	7.245	8:00 am—9:00 am
Nouakchott	9.610	8:00 am—9:00 am
MAURITIUS		
Forest Side	4.850	8:00 am—Noon
MEXICO		
Mexico City	11.770	9:00 am—Noon
Mexico City	17.835	9:00 am—Noon
Mexico City	21.705	9:00 am—Noon
MONACO		
Monte Carlo	6.035	8:00 am—Noon
Monte Carlo	7.135	8:00 am—Noon

Location	Freq (MHz)	Actual Time
MOZAMBIQUE		
Lourenco Marques	6.050	8:00 am–11:00 am
Lourenco Marques	9.620	8:00 am–9:00 am
Lourenco Marques	11.780	8:00 am–Noon
NETHERLANDS ANTILLES		
Bonaire	15.255	8:00 am–10:00 am
NICARAGUA		
Managua	11.875	8:00 am–Noon
NIGER		
Niamey	7.155	8:00 am–8:30 am
NIGERIA		
Lagos	7.275	10:30 am–Noon
Lagos	15.120	10:30 am–Noon
Lagos	15.185	10:30 am–Noon
NORWAY		
Oslo	21.655	10:00 am–10:30 am
PAKISTAN		
Karachi	21.590	8:00 am–9:00 am
PANAMA		
David	6.045	8:00 am–Noon
PARAGUAY		
Asunción	5.975	8:00 am–Noon
Asunción	11.945	8:00 am–Noon
Asunción	15.210	8:00 am–Noon
PHILIPPINES		
Manila	11.920	8:00 am–9:00 am
PORTUGAL		
Lisbon	17.895	8:45 am–9:30 am
Lisbon	21.495	8:45 am–9:30 am
REUNION		
St. Denis	4.807	8:00 am–Noon

Location	Freq (MHz)	Actual Time
RHODESIA		
Salisbury	3.396	10:45 am–Noon
Salisbury	7.285	8:00 am–10:45 am
ROMANIA		
Bucharest	11.920	10:00 am–10:30 am
Bucharest	15.250	8:00 am–8:30 am
		10:00 am–10:30 am
Bucharest	17.710	8:00 am–8:30 am
RWANDA		
Kigali	6.055	8:30 am–Noon
SARAWAK		
Kuching	7.160	9:00 am–9:45 am
SEYCHELLES		
Victoria	11.930	8:30 am–9:30 am
		10:30 am–11:45 am
Victoria	15.270	8:30 am–9:30 am
		10:30 am–11:45 am
SINGAPORE		
Singapore	5.052	8:00 am–11:30 am
Singapore	11.940	8:00 am–11:30 am
SOUTH AFRICA		
Johannesburg	11.900	8:00 am–10:00 am
		11:00 am–Noon
Johannesburg	15.220	8:00 am–Noon
Johannesburg	21.535	8:00 am–Noon
SRI LANKA (Ceylon)		
Colombo	4.968	8:00 am–Noon
Colombo	5.020	8:00 am–Noon
SWEDEN		
Stockholm	6.065	11:00 am–11:30 am
Stockholm	11.930	11:00 am–11:30 am
Stockholm	15.240	9:00 am–9:30 am
Stockholm	21.505	9:00 am–9:30 am

Location	Freq (MHz)	Actual Time
SWITZERLAND		
Berne	11.865	10:15 am–10:45 am
Berne	15.305	8:15 am–8:45 am
		10:15 am–10:45 am
Berne	17.830	10:15 am–10:45 am
Berne	17.845	8:15 am–8:45 am
Berne	21.520	8:15 am–8:45 am
Berne	21.605	8:15 am–8:45 am
TANZANIA		
Dar es Salaam	4.785	11:00 am–Noon
Dar es Salaam	6.105	11:00 am–Noon
TURKEY		
Ankara	17.820	8:30 am–9:00 am
UGANDA		
Kampala	7.110	9:00 am–9:30 am
Kampala	7.195	9:00 am–9:30 am
UNITED ARAB REPUBLIC (Egypt)		
Cairo	17.920	8:15 am–9:30 am
UNITED STATES OF AMERICA		
Washington, D. C. (VOA)	3.980	11:00 am–Noon
Washington, D. C. (VOA)	6.040	11:00 am–Noon
Washington, D. C. (VOA)	11.790	9:00 am–Noon
Washington, D. C. (VOA)	15.205	11:00 am–Noon
Washington, D. C. (VOA)	17.785	11:00 am–Noon
Washington, D. C. (VOA)	21.485	9:00 am–Noon
Washington, D. C. (VOA)	21.650	9:00 am–Noon
UPPER VOLTA		
Ouagadougou	9.515	8:00 am–9:00 am
U.S.S.R.		
Moscow	15.230	11:00 am–Noon
Moscow	15.295	10:00 am–11:00 am
Moscow	17.775	11:00 am–Noon
VATICAN STATE		
Vatican City	9.615	8:00 am–8:15 am

Location	Freq (MHz)	Actual Time
VATICAN STATE (cont)		
Vatican City	9.645	10:00 am—10:30 am
Vatican City	11.705	11:20 am—Noon
Vatican City	11.725	8:00 am—8:15 am
Vatican City	11.740	10:00 am—10:30 am
Vatican City	15.120	10:00 am—10:30 am
Vatican City	15.155	8:00 am—8:15 am
Vatican City	15.210	11:30 am—Noon
VIETNAM (Democratic Rep.)		
Hanoi	7.470	8:30 am—9:00 am
		9:30 am—10:00 am
		10:30 am—11:00 am
Hanoi	10.010	8:30 am—9:00 am
		9:30 am—10:00 am
		10:30 am—11:00 am
Hanoi	12.025	8:00 am—8:30 am
Hanoi	15.004	8:00 am—8:30 am
YUGOSLAVIA		
Belgrade	9.620	10:30 am—11:00 am
Belgrade	11.735	10:30 am—11:00 am
Belgrade	15.240	10:30 am—11:00 am
ZAIRE		
Kinshasa	7.115	8:00 am—Noon
Kinshasa	15.265	8:00 am—Noon
ZAMBIA		
Lusaka	6.165	8:00 am—Noon
Lusaka	7.240	8:00 am—Noon

Part 4—Noon—4:00 pm EST

AFGHANISTAN		
Kabul	9.530	1:00 pm—1:30 pm
Kabul	11.790	1:00 pm—1:30 pm
Kabul	15.265	1:00 pm—1:30 pm
Kabul	17.780	1:00 pm—1:30 pm

Location	Freq (MHz)	Actual Time
ALGERIA		
Algiers	11.715	Noon–4:00 pm
Algiers	11.835	Noon–4:00 pm
Algiers	15.420	Noon–4:00 pm
ANGOLA		
Luanda	7.245	Noon–4:00 pm
Luanda	9.535	Noon–2:00 pm
Luanda	11.875	Noon–2:00 pm
ARGENTINA		
Buenos Aires	9.760	1:00 pm–4:00 pm
Buenos Aires	11.880	Noon–4:00 pm
BANGLADESH		
Dacca	11.650	12:15 pm–1:00 pm
BOLIVIA		
La Paz	6.005	Noon–4:00 pm
La Paz	9.555	Noon–4:00 pm
BRAZIL		
Brasilia	6.065	Noon–4:00 pm
Rio de Janeiro	9.720	Noon–4:00 pm
Rio de Janeiro	11.795	Noon–4:00 pm
Sao Paulo	6.125	Noon–4:00 pm
BULGARIA		
Sofia	6.070	2:30 pm–3:00 pm
Sofia	9.700	2:30 pm–3:00 pm
CAMEROON		
Yaoundé	4.972	12:30 pm–1:45 pm
Yaoundé	9.760	12:30 pm–1:45 pm
CANADA		
Montreal	15.325	1:30 pm–2:15 pm
Montreal	17.820	1:30 pm–2:15 pm
CANARY ISLANDS		
Santa Cruz	11.800	3:00 pm–4:00 pm
Santa Cruz	15.365	3:00 pm–4:00 pm

Location	Freq (MHz)	Actual Time
CAPE VERDE ISLANDS		
Sao Vicente	4.715	1:00 pm–3:00 pm
CENTRAL AFRICAN REPUBLIC		
Bangui	5.035	Noon–4:00 pm
CHAD		
Fort Lamy	4.905	Noon–4:00 pm
CHINA (People's Rep. of)		
Peking	7.590	3:30 pm–4:00 pm
Peking	9.030	3:30 pm–4:00 pm
Peking	11.650	3:30 pm–4:00 pm
CHINA (Rep. of) (Taiwan)		
Taipei	9.685	1:00 pm–2:00 pm
Taipei	9.765	1:00 pm–2:00 pm
Taipei	11.825	1:00 pm–2:00 pm
Taipei	15.125	1:00 pm–2:00 pm
Taipei	15.370	1:00 pm–2:00 pm
Taipei	17.720	1:00 pm–2:00 pm
Taipei	17.890	1:00 pm–2:00 pm
COLOMBIA		
Bogotá	6.030	Noon–4:00 pm
Bogotá	6.183	Noon–4:00 pm
Bogotá	11.825	Noon–4:00 pm
COMORO ISLANDS		
Moroni	3.331	Noon–2:30 pm
CONGO (Republic of)		
Brazzaville	15.190	3:15 pm–4:00 pm
COOK ISLANDS		
Rarotonga	5.045	Noon–1:30 pm
COSTA RICA		
San José	6.075	Noon–4:00 pm
CUBA		
Havana	15.155	3:15 pm–4:00 pm

Location	Freq (MHz)	Actual Time
CUBA (cont)		
Havana	15.270	3:50 pm–4:00 pm
Havana	15.285	3:50 pm–4:00 pm
CZECHOSLOVAKIA		
Prague	5.930	12:30 pm–1:30 pm
		2:00 pm–2:30 pm
Prague	7.345	12:30 pm–1:30 pm
		2:00 pm–2:30 pm
Prague	17.840	12:30 pm–1:30 pm
DAHOMEY		
Cotonou	3.270	Noon–1:00 pm
		2:00 pm–2:45 pm
Cotonou	4.870	Noon–1:00 pm
		2:00 pm–2:45 pm
DOMINICAN REPUBLIC		
Santo Domingo	6.090	Noon–4:00 pm
Santo Domingo	9.505	Noon–4:00 pm
ECUADOR		
Quito	15.300	2:00 pm–3:15 pm
Quito	21.460	2:00 pm–3:15 pm
ELLICE ISLAND		
Ellice	3.230	1:45 pm–4:00 pm
ETHIOPIA		
Addis Ababa	7.145	1:30 pm–2:15 pm
Addis Ababa	11.910	2:30 pm–3:15 pm
FIJI ISLANDS		
Suva	3.230	1:00 pm–4:00 pm
FINLAND		
Helsinki	15.185	Noon–1:30 pm
FRANCE		
Paris	15.295	3:15 pm–4:00 pm
Paris	17.720	3:15 pm–4:00 pm
Paris	21.580	3:15 pm–4:00 pm

Location	Freq (MHz)	Actual Time
GABON		
Libreville	4.777	Noon–4:00 pm
GAMBIA		
Bathurst	4.820	Noon–4:00 pm
GERMANY (Democratic Rep.)		
Berlin	15.390	1:15 pm–2:00 pm
Berlin	21.475	1:15 pm–2:00 pm
GHANA		
Accra	6.130	Noon–4:00 pm
Accra	9.545	3:45 pm–4:00 pm
Accra	11.850	3:00 pm–4:00 pm
Accra	15.285	1:15 pm–2:00 pm
GREAT BRITAIN		
London	5.975	Noon–4:00 pm
London	6.180	Noon–4:00 pm
London	7.120	1:00 pm–4:00 pm
London	9.410	Noon–4:00 pm
London	12.095	Noon–1:00 pm
London	15.070	Noon–1:00 pm
GREECE		
Athens	7.295	1:30 pm–2:00 pm
Athens	9.605	1:30 pm–2:00 pm
		2:45 pm–3:00 pm
Athens	11.720	2:45 pm–3:00 pm
GUATEMALA		
Guatemala City	6.180	Noon–4:00 pm
Guatemala City	9.505	Noon–4:00 pm
GUINEA (Portuguese)		
Bissau	5.045	Noon–4:00 pm
GUINEA (Republic)		
Conakry	7.125	Noon–4:00 pm
Conakry	9.650	Noon–4:00 pm
Conakry	15.310	Noon–4:00 pm

Location	Freq (MHz)	Actual Time
GUIANA (French)		
Cayenne	3.385	Noon–1:00 pm
		3:30 pm–4:00 pm
GUYANA		
Georgetown	3.265	Noon–4:00 pm
HOLLAND		
Hilversum	6.020	1:30 pm–3:00 pm
Hilversum	6.085	1:30 pm–3:00 pm
Hilversum	17.830	1:30 pm–3:00 pm
HONDURAS (Republic)		
Tegucigalpa	4.820	Noon–4:00 pm
Tegucipalpa	6.050	Noon–4:00 pm
HUNGARY		
Budapest	7.220	2:30 pm–3:00 pm
Budapest	9.833	2:30 pm–3:00 pm
Budapest	11.910	2:30 pm–3:00 pm
INDIA		
New Delhi	9.690	2:45 pm–3:45 pm
New Delhi	9.912	2:45 pm–4:00 pm
New Delhi	11.620	12:45 pm–2:45 pm
New Delhi	11.945	12:45 pm–2:45 pm
New Delhi	11.960	2:45 pm–3:45 pm
New Delhi	15.080	12:45 pm–2:45 pm
INDONESIA		
Jakarta	9.585	2:00 pm–3:00 pm
Jakarta	11.795	2:00 pm–3:00 pm
IRAN		
Tehran	12.165	12:30 pm–3:30 pm
Tehran	15.084	12:30 pm–3:30 pm
IRAQ		
Baghdad	9.745	2:30 pm–3:30 pm
ITALY		
Rome	7.235	3:30 pm–3:45 pm

Location	Freq (MHz)	Actual Time
ITALY (cont)		
Rome	7.275	2:30 pm–3:00 pm
Rome	9.575	3:30 pm–3:45 pm
Rome	9.710	2:30 pm–3:00 pm
Rome	11.810	2:30 pm–3:00 pm
		3:30 pm–3:45 pm
IVORY COAST		
Abidjan	7.215	1:00 pm–4:00 pm
Abidjan	11.920	1:00 pm–4:00 pm
JAPAN		
Tokyo	9.505	Noon–12:15 pm
		1:00 pm–1:15 pm
Tokyo	11.815	Noon–12:15 pm
		1:00 pm–1:15 pm
		2:00 pm–2:15 pm
		3:00 pm–3:15 pm
Tokyo	15.105	2:00 pm–2:15 pm
		3:00 pm–3:15 pm
Tokyo	15.390	Noon–12:15 pm
		1:00 pm–1:15 pm
		2:00 pm–2:15 pm
		3:00 pm–3:15 pm
KENYA		
Nairobi	4.915	Noon–3:00 pm
KOREA (Dem. People's Rep.)		
Pyongyang	6.540	2:00 pm–3:00 pm
Pyongyang	9.515	2:00 pm–3:00 pm
KUWAIT		
Kuwait	9.595	Noon–1:00 pm
Kuwait	11.925	1:30 pm–4:00 pm
Kuwait	15.345	Noon–2:00 pm
LEBANON		
Beirut	11.705	1:15 pm–3:30 pm
LESOTHO		
Maseru	4.800	Noon–3:00 pm

Location	Freq (MHz)	Actual Time
LIBERIA		
Monrovia	4.770	Noon–2:00 pm
Monrovia	11.940	2:15 pm–4:00 pm
LUXEMBOURG		
Villa Louvigny	6.090	1:30 pm–4:00 pm
MALAWI		
Zomba	3.380	1:00 pm–4:00 pm
Zomba	5.995	1:00 pm–4:00 pm
MALI		
Bamako	17.725	Noon–12:30 pm
MAURITANIA		
Nouakchott	4.850	Noon–4:00 pm
MAURITIUS		
Forest Side	4.850	Noon–1:30 pm
MEXICO		
Mexico City	11.770	Noon–4:00 pm
Mexico City	17.835	Noon–4:00 pm
Mexico City	21.705	Noon–4:00 pm
MONACO		
Monte Carlo	6.035	Noon–4:00 pm
Monte Carlo	7.135	Noon–4:00 pm
MOROCCO		
Tangier	7.225	Noon–1:00 pm
Tangier	11.735	Noon–1:00 pm
MOZAMBIQUE		
Lourenco Marques	11.780	Noon–2:00 pm
NEW CALEDONIA		
Noumea	3.355	2:00 pm–4:00 pm
Noumea	7.170	2:00 pm–4:00 pm
NEW ZEALAND		
Wellington	11.780	Noon–3:00 pm

Location	Freq (MHz)	Actual Time
NEW ZEALAND (cont)		
Wellington	15.110	3:00 pm–4:00 pm
Wellington	17.770	3:00 pm–4:00 pm
NICARAGUA		
Managua	11.875	Noon–4:00 pm
NIGERIA		
Lagos	7.275	1:00 pm–2:30 pm
Lagos	11.925	1:00 pm–2:30 pm
Lagos	15.185	1:00 pm–2:30 pm
NORWAY		
Oslo	11.850	Noon–12:30 pm
		2:00 pm–2:30 pm
Oslo	21.655	2:00 pm–2:30 pm
PAKISTAN		
Karachi	7.095	1:45 pm–4:00 pm
Karachi	9.460	1:45 pm–4:00 pm
PANAMA		
David	6.045	Noon–4:00 pm
PARAGUAY		
Asunción	5.975	Noon–1:00 pm
Asunción	15.210	Noon–4:00 pm
PORTUGAL		
Lisbon	6.025	3:45 pm–4:00 pm
Lisbon	11.935	1:15 pm–2:15 pm
Lisbon	21.495	1:15 pm–2:15 pm
REUNION		
St. Denis	4.807	Noon–1:45 pm
RHODESIA		
Salisbury	3.396	Noon–4:00 pm
ROMANIA		
Bucharest	6.150	2:30 pm–3:00 pm
Bucharest	7.195	2:30 pm–3:30 pm

Location	Freq (MHz)	Actual Time
RWANDA		
Kigali	6.055	Noon–4:00 pm
SAUDI ARABIA		
Mecca	11.855	Noon–3:00 pm
SENEGAL		
Dakar	4.890	1:15 pm–1:45 pm
Dakar	11.895	1:15 pm–1:45 pm
SOLOMON ISLANDS		
Honiara	9.775	2:00 pm–4:00 pm
SOMALIA		
Mogadishu	9.588	12:30 pm–1:00 pm
SOUTH AFRICA		
Johannesburg	9.695	2:00 pm–3:00 pm
Johannesburg	15.155	1:00 pm–2:00 pm
Johannesburg	15.175	2:00 pm–3:00 pm
Johannesburg	21.480	1:00 pm–2:00 pm
SUDAN		
Omdurman	6.150	12:15 pm–1:00 pm
Omdurman	7.200	12:15 pm–1:00 pm
Omdurman	9.505	12:15 pm–1:00 pm
SWEDEN		
Stockholm	6.065	3:45 pm–4:00 pm
Stockholm	9.745	3:45 pm–4:00 pm
Stockholm	11.950	1:30 pm–2:00 pm
Stockholm	15.240	1:30 pm–2:00 pm
TANZANIA		
Dar es Salaam	4.785	Noon–1:30 pm
Dar es Salaam	6.105	Noon–1:15 pm
Dar es Salaam	15.435	1:30 pm–2:45 pm
TOGO		
Lomé	5.047	2:45 pm–3:00 pm
Lomé	7.265	2:45 pm–3:00 pm

Location	Freq (MHz)	Actual Time
TURKS & CAICOS ISLANDS		
Grand Turk	4.788	2:30 pm–3:00 pm
UNITED ARAB REPUBLIC (Egypt)		
Cairo	17.655	12:30 pm–1:45 pm
Cairo	17.725	3:30 pm–4:00 pm
UNITED STATES OF AMERICA		
New York, N.Y.	9.690	3:00 pm–4:00 pm
New York, N.Y.	11.890	3:00 pm–4:00 pm
New York, N.Y.	15.215	2:00 pm–2:45 pm
New York, N.Y.	17.845	Noon–2:45 pm
New York, N.Y.	21.525	Noon–1:45 pm
Washington, D.C. (VOA)	3.980	Noon–4:00 pm
Washington, D.C. (VOA)	6.040	Noon–4:00 pm
Washington, D.C. (VOA)	6.160	1:00 pm–4:00 pm
Washington, D.C. (VOA)	7.285	1:00 pm–4:00 pm
Washington, D.C. (VOA)	9.760	1:00 pm–4:00 pm
Washington, D.C. (VOA)	11.790	Noon–4:00 pm
Washington, D.C. (VOA)	15.205	Noon–4:00 pm
Washington, D.C. (VOA)	17.785	Noon–4:00 pm
Washington, D.C. (VOA)	21.485	Noon–4:00 pm
Washington, D.C. (VOA)	21.650	Noon–4:00 pm
UPPER VOLTA		
Ouagadougou	9.515	Noon–4:00 pm
VATICAN STATE		
Vatican City	6.190	3:00 pm–4:00 pm
Vatican City	7.250	3:00 pm–4:00 pm
Vatican City	9.645	3:00 pm–4:00 pm
Vatican City	11.705	Noon–3:00 pm
Vatican City	11.740	3:45 pm–4:00 pm
Vatican City	15.120	3:45 pm–4:00 pm
Vatican City	15.210	Noon–3:00 pm
VIETNAM (Democratic Rep.)		
Hanoi	12.025	3:00 pm–3:30 pm
Hanoi	15.004	3:00 pm–3:30 pm

Location	Freq (MHz)	Actual Time
YUGOSLAVIA		
Belgrade	6.100	1:30 pm–2:00 pm
		3:00 pm–3:30 pm
Belgrade	7.240	1:30 pm–2:00 pm
		3:00 pm–3:30 pm
Belgrade	9.620	1:30 pm–2:00 pm
		3:00 pm–3:30 pm
ZAIRE		
Kinshasa	7.115	Noon–4:00 pm
Kinshasa	15.265	Noon–4:00 pm
ZAMBIA		
Lusaka	6.165	Noon–3:00 pm
Lusaka	7.240	Noon–3:00 pm

Part 5—4:00 pm–8:00 pm EST

ALBANIA		
Tirana	7.065	7:00 pm–7:30 pm
Tirana	9.780	7:00 pm–7:30 pm
ALGERIA		
Algiers	11.715	4:00 pm–7:00 pm
Algiers	11.835	4:00 pm–7:00 pm
Algiers	15.420	4:00 pm–7:00 pm
ANGOLA		
Luanda	7.245	4:00 pm–7:00 pm
ARGENTINA		
Buenos Aires	9.760	4:00 pm–8:00 pm
Buenos Aires	11.880	4:00 pm–8:00 pm
AUSTRALIA		
Melbourne	6.055	4:00 pm–5:15 pm
Melbourne	11.840	6:30 pm–7:15 pm
Melbourne	17.715	7:30 pm–8:00 pm

Location	Freq (MHz)	Actual Time
AUSTRIA		
Vienna	9.770	6:00 pm–7:00 pm
Vienna	15.145	7:00 pm–8:00 pm
BELGIUM		
Brussels	9.550	6:05 pm–6:15 pm
		7:50 pm–8:00 pm
Brussels	11.875	6:05 pm–6:15 pm
		7:50 pm–8:00 pm
BOLIVIA		
La Paz	6.005	4:00 pm–8:00 pm
La Paz	9.555	4:00 pm–8:00 pm
BRAZIL		
Brasilia	6.065	4:00 pm–8:00 pm
Rio de Janeiro	9.720	4:00 pm–8:00 pm
Rio de Janeiro	11.795	4:00 pm–8:00 pm
Sao Paulo	6.125	4:00 pm–8:00 pm
BRUNEI		
Tutong	7.215	6:00 pm–7:30 pm
BULGARIA		
Sofia	6.070	4:30 pm–5:00 pm
Sofia	9.700	4:30 pm–5:00 pm
		7:00 pm–8:00 pm
CAMBODIA		
Phnom Penh	4.907	6:30 pm–7:00 pm
		7:45 pm–8:00 pm
CANADA		
Montreal	9.625	6:00 pm–6:30 pm
Montreal	11.945	6:00 pm–6:30 pm
Montreal	15.190	6:00 pm–6:30 pm
Montreal	15.325	4:15 pm–5:00 pm
CANARY ISLANDS		
Santa Cruz	11.800	4:00 pm–8:00 pm
Santa Cruz	15.365	4:00 pm–8:00 pm

Location	Freq (MHz)	Actual Time
CAPE VERDE ISLANDS		
Sao Vicente	3.910	5:00 pm–8:00 pm
CENTRAL AFRICAN REPUBLIC		
Bangui	5.035	4:00 pm–6:00 pm
CHAD		
Fort Lamy	4.905	4:00 pm–4:30 pm
CHINA (People's Rep. of)		
Peking	7.590	4:00 pm–5:30 pm
Peking	9.030	4:00 pm–5:30 pm
Peking	11.650	4:00 pm–5:30 pm
Peking	11.675	7:00 pm–8:00 pm
Peking	15.060	7:00 pm–8:00 pm
COLOMBIA		
Bogotá	6.030	4:00 pm–8:00 pm
Bogotá	6.183	4:00 pm–8:00 pm
Bogotá	11.825	4:00 pm–8:00 pm
COOK ISLANDS		
Rarotonga	9.695	5:45 pm–6:00 pm
COSTA RICA		
San José	6.075	4:00 pm–8:00 pm
CUBA		
Havana	15.155	4:00 pm–4:45 pm
Havana	15.270	4:00 pm–4:50 pm
Havana	15.285	4:00 pm–4:50 pm
DOMINICAN REPUBLIC		
Santo Domingo	6.090	4:00 pm–8:00 pm
Santo Domingo	9.505	4:00 pm–8:00 pm
ECUADOR		
Quito	11.745	6:30 pm–7:00 pm
EL SALVADOR		
San Salvador	5.980	5:00 pm–8:00 pm
San Salvador	9.555	5:00 pm–8:00 pm

Location	Freq (MHz)	Actual Time
FIJI ISLANDS		
Suva	3.230	4:00 pm–4:15 pm
Suva	6.005	4:15 pm–8:00 pm
GABON		
Libreville	4.777	4:00 pm–6:00 pm
GAMBIA		
Bathurst	4.820	4:00 pm–6:00 pm
GHANA		
Accra	6.130	4:00 pm–4:15 pm
Accra	9.545	4:00 pm–5:15 pm
GREAT BRITAIN		
London	5.975	4:00 pm–4:15 pm
London	6.110	4:15 pm–8:00 pm
London	6.180	4:00 pm–7:30 pm
London	7.120	4:00 pm–8:00 pm
London	7.130	5:00 pm–8:00 pm
London	9.410	4:00 pm–6:00 pm
London	9.510	7:30 pm–8:00 pm
London	11.780	4:15 pm–6:15 pm
GREECE		
Athens	9.605	5:15 pm–6:00 pm
Athens	11.720	5:15 pm–6:30 pm
GUATEMALA		
Guatemala City	6.180	4:00 pm–8:00 pm
Guatemala City	9.505	4:00 pm–8:00 pm
GUINEA (Portuguese)		
Bissau	5.045	4:00 pm–7:00 pm
GUINEA (Republic)		
Conakry	7.125	4:00 pm–7:00 pm
Conakry	9.650	4:00 pm–6:00 pm
Conakry	15.310	4:00 pm–7:00 pm
GUIANA (French)		
Cayenne	3.385	4:00 pm–8:00 pm

Location	Freq (MHz)	Actual Time
GUYANA		
Georgetown	3.265	4:00 pm–8:00 pm
HAITI		
Cap Haitien	9.770	4:00 pm–8:00 pm
Cap Haitien	11.835	4:00 pm–8:00 pm
HOLLAND		
Hilversum	7.290	4:30 pm–6:00 pm
Hilversum	9.715	4:30 pm–6:00 pm
Hilversum	9.740	4:30 pm–6:00 pm
HONDURAS (British)		
Belize	3.300	5:00 pm–8:00 pm
HONDURAS (Republic)		
Tegucigalpa	4.820	4:00 pm–8:00 pm
Tegucigalpa	6.050	4:00 pm–8:00 pm
HUNGARY		
Budapest	7.220	4:30 pm–5:00 pm
Budapest	9.833	4:30 pm–5:00 pm
Budapest	11.910	4:30 pm–5:00 pm
INDIA		
New Delhi	7.235	5:45 pm–7:00 pm
New Delhi	9.570	5:45 pm–7:00 pm
New Delhi	9.912	4:00 pm–5:30 pm
New Delhi	11.710	5:45 pm–7:00 pm
New Delhi	15.235	7:15 pm–8:00 pm
INDONESIA		
Jakarta	9.585	6:30 pm–7:00 pm
Jakarta	11.795	6:30 pm–7:00 pm
ISRAEL		
Jerusalem	9.009	4:30 pm–5:00 pm
Jerusalem	9.625	4:30 pm–5:00 pm
Jerusalem	9.725	4:30 pm–5:00 pm
ITALY		
Rome	5.990	5:00 pm–5:30 pm

Location	Freq (MHz)	Actual Time
ITALY (cont)		
Rome	9.575	5:00 pm–5:30 pm
IVORY COAST		
Abidjan	7.215	4:00 pm–7:00 pm
Abidjan	11.920	4:00 pm–7:00 pm
JAPAN		
Tokyo	11.815	4:00 pm–4:15 pm
		5:00 pm–5:15 pm
		6:00 pm–6:15 pm
Tokyo	15.105	4:00 pm–4:15 pm
Tokyo	15.390	4:00 pm–4:15 pm
		5:00 pm–5:15 pm
		6:00 pm–6:15 pm
Tokyo	17.785	5:00 pm–5:15 pm
		6:00 pm–6:15 pm
KOREA (Republic of)		
Seoul	9.640	4:00 pm–4:30 pm
Seoul	15.430	4:00 pm–4:30 pm
LUXEMBOURG		
Villa Louvigny	6.090	4:00 pm–8:00 pm
MALAWI		
Zomba	3.380	4:00 pm–5:00 pm
Zomba	5.995	4:00 pm–5:00 pm
MALAYSIA		
Kuala Lumpur	9.660	5:30 pm–8:00 pm
Kuala Lumpur	11.900	5:30 pm–8:00 pm
Kuala Lumpur	15.280	5:30 pm–8:00 pm
MALI		
Bamako	17.725	4:00 pm–8:00 pm
MAURITANIA		
Nouakchott	4.850	4:00 pm–6:00 pm
MEXICO		
Mexico City	11.770	4:00 pm–8:00 pm

Location	Freq (MHz)	Actual Time
MEXICO (cont)		
Mexico City	17.835	4:00 pm–8:00 pm
Mexico City	21.705	4:00 pm–8:00 pm
MONACO		
Monte Carlo	6.035	4:00 pm–8:00 pm
Monte Carlo	7.135	4:00 pm–8:00 pm
MONGOLIA		
Ulan Bator	9.540	5:00 pm–5:30 pm
Ulan Bator	11.860	5:00 pm–5:30 pm
NETHERLANDS ANTILLES		
Bonaire	11.820	7:30 pm–8:00 pm
NEW CALEDONIA		
Noumea	3.355	6:30 pm–8:00 pm
Noumea	7.170	6:30 pm–8:00 pm
NEW GUINEA		
Port Moresby	9.520	5:15 pm–8:00 pm
NEW ZEALAND		
Wellington	15.110	4:00 pm–8:00 pm
Wellington	17.770	4:00 pm–8:00 pm
NICARAGUA		
Managua	11.875	4:00 pm–8:00 pm
NORWAY		
Oslo	9.645	6:00 pm–6:30 pm
Oslo	11.850	4:00 pm–4:30 pm
PAKISTAN		
Karachi	7.095	4:00 pm–4:30 pm
Karachi	7.265	7:15 pm–8:00 pm
Karachi	9.460	4:00 pm–4:30 pm
PANAMA		
David	6.045	4:00 pm–8:00 pm

Location	Freq (MHz)	Actual Time
PARAGUAY		
Asunción	5.975	4:00 pm–8:00 pm
Asunción	15.210	4:00 pm–8:00 pm
PHILIPPINES		
Manila	15.235	6:30 pm–7:00 pm
Manila	21.515	7:00 pm–8:00 pm
PORTUGAL		
Lisbon	6.025	4:00 pm–4:30 pm
ROMANIA		
Bucharest	6.190	4:00 pm–4:30 pm
Bucharest	7.225	4:00 pm–4:30 pm
SARAWAK		
Kuching	7.160	6:30 pm–8:00 pm
SINGAPORE		
Singapore	5.052	5:30 pm–8:00 pm
Singapore	11.940	5:30 pm–8:00 pm
SOLOMON ISLANDS		
Honiara	9.775	4:00 pm–7:00 pm
SOUTH AFRICA		
Johannesburg	6.080	6:30 pm–8:00 pm
Johannesburg	9.525	5:15 pm–6:15 pm
Johannesburg	9.560	6:30 pm–8:00 pm
Johannesburg	9.695	5:15 pm–8:00 pm
Johannesburg	11.900	5:15 pm–6:15 pm
Johannesburg	11.970	5:15 pm–8:00 pm
SWEDEN		
Stockholm	6.065	4:00 pm–4:15 pm
Stockholm	6.175	7:30 pm–8:00 pm
Stockholm	9.745	4:00 pm–4:15 pm
SWITZERLAND		
Berne	9.590	4:00 pm–4:45 pm
Berne	11.720	4:00 pm–4:45 pm

Location	Freq (MHz)	Actual Time
SWITZERLAND (cont)		
Berne	11.865	4:00 pm–4:45 pm
Berne	15.305	4:00 pm–4:45 pm
TURKEY		
Ankara	9.515	5:00 pm–5:30 pm
Ankara	15.195	5:00 pm–5:30 pm
UNITED ARAB REPUBLIC (Egypt)		
Cairo	9.850	4:45 pm–6:00 pm
Cairo	17.725	4:00 pm–5:00 pm
UNITED STATES OF AMERICA		
New York, N.Y.	9.690	4:00 pm–6:00 pm
New York, N.Y.	11.890	4:00 pm–5:15 pm
Washington, D.C. (VOA)	3.980	4:00 pm–7:00 pm
Washington, D.C. (VOA)	6.040	4:00 pm–7:00 pm
Washington, D.C. (VOA)	6.160	4:00 pm–7:00 pm
Washington, D.C. (VOA)	7.285	4:00 pm–7:00 pm
Washington, D.C. (VOA)	9.760	4:00 pm–7:00 pm
Washington, D.C. (VOA)	11.760	5:00 pm–7:00 pm
Washington, D.C. (VOA)	11.790	4:00 pm–6:15 pm
Washington, D.C. (VOA)	15.205	4:00 pm–7:00 pm
Washington, D.C. (VOA)	21.485	4:00 pm–6:15 pm
Washington, D.C. (VOA)	21.650	4:00 pm–6:15 pm
UPPER VOLTA		
Ouagadougou	9.515	4:00 pm–6:00 pm
U. S. S. R		
Moscow	9.660	5:00 pm–5:30 pm
		6:00 pm–6:30 pm
		7:00 pm–7:30 pm
Moscow	9.668	5:00 pm–5:30 pm
		6:00 pm–6:30 pm
		7:00 pm–7:30 pm
Moscow	11.850	5:00 pm–5:30 pm
		6:00 pm–6:30 pm
		7:00 pm–7:30 pm
Moscow	11.870	5:00 pm–5:30 pm
		6:00 pm–6:30 pm
		7:00 pm–7:30 pm

Location	Freq (MHz)	Actual Time
U.S.S.R. (cont)		
Moscow	11.900	5:00 pm–5:30 pm
		6:00 pm–6:30 pm
		7:00 pm–7:30 pm
Moscow	11.920	5:00 pm–5:30 pm
		6:00 pm–6:30 pm
		7:00 pm–7:30 pm
Moscow	15.210	5:00 pm–5:30 pm
		6:00 pm–6:30 pm
		7:00 pm–7:30 pm
Moscow	17.730	5:00 pm–5:30 pm
		6:00 pm–6:30 pm
		7:00 pm–7:30 pm
VIETNAM (Democratic Rep.)		
Hanoi	10.010	6:30 pm–7:00 pm
WINDWARD ISLANDS		
Grenada	11.975	7:00 pm–8:00 pm
YUGOSLAVIA		
Belgrade	6.100	5:00 pm–5:15 pm
Belgrade	7.240	5:00 pm–5:15 pm
Belgrade	9.620	5:00 pm–5:15 pm
ZAIRE		
Kinshasa	7.115	4:00 pm–8:00 pm
Kinshasa	15.265	4:00 pm–8:00 pm

Part 6—8:00 pm—Midnight EST

Location	Freq (MHz)	Actual Time
AFARS & ISSAS		
Djibouti	4.780	10:00 pm–Midnight
ALBANIA		
Tirana	6.195	8:30 pm–9:00 pm
		9:30 pm–10:00 pm
		10:30 pm–11:00 pm

Location	Freq (MHz)	Actual Time
ALBANIA (cont)		
Tirana	7.300	8:30 pm–9:00 pm
		9:30 pm–10:00 pm
		10:30 pm–11:00 pm
ARGENTINA		
Buenos Aires	9.760	8:00 pm–Midnight
Buenos Aires	11.880	8:00 pm–11:00 pm
AUSTRALIA		
Melbourne	15.320	8:00 pm–10:00 pm
Melbourne	17.715	8:00 pm–9:00 pm
		11:00 pm–Midnight
Melbourne	17.795	8:00 pm–10:00 pm
Melbourne	21.740	8:00 pm–10:00 pm
AUSTRIA		
Vienna	6.115	11:00 pm–Midnight
Vienna	15.145	8:00 pm–9:00 pm
BOLIVIA		
La Paz	6.005	8:00 pm–11:30 pm
La Paz	9.555	8:00 pm–11:00 pm
BOTSWANA		
Gaberones	3.356	11:15 pm–11:30 pm
Gaberones	4.845	11:15 pm–11:30 pm
BRAZIL		
Brasilia	6.065	8:00 pm–Midnight
Rio de Janeiro	9.720	8:00 pm–11:00 pm
Rio de Janeiro	11.795	8:00 pm–10:00 pm
Sao Paulo	6.125	8:00 pm–Midnight
BRUNEI		
Tutong	7.215	10:00 pm–Midnight
BULGARIA		
Sofia	9.700	11:00 pm–Midnight
BURMA		
Rangoon	7.120	9:00 pm–9:30 pm

Location	Freq (MHz)	Actual Time
CANARY ISLANDS		
Santa Cruz	11.800	8:00 pm–Midnight
Santa Cruz	15.365	8:00 pm–Midnight
CHAD		
Fort Lamy	4.905	11:30 pm–Midnight
CHINA (People's Rep. of)		
Peking	7.120	8:00 pm–11:00 pm
Peking	9.780	8:00 pm–9:00 pm
		10:00 pm–11:00 pm
Peking	11.675	8:00 pm–Midnight
Peking	15.060	8:00 pm–Midnight
Peking	15.095	10:00 pm–Midnight
Peking	15.385	10:00 pm–Midnight
Peking	17.715	8:00 pm–10:00 pm
Peking	17.735	10:00 pm–Midnight
CHINA (Rep. of) (Taiwan)		
Taipei	7.130	9:00 pm–11:00 pm
Taipei	11.825	9:00 pm–11:00 pm
Taipei	15.125	9:00 pm–11:00 pm
Taipei	15.345	9:00 pm–11:00 pm
Taipei	17.720	9:00 pm–11:00 pm
Taipei	17.780	9:00 pm–11:00 pm
Taipei	17.890	9:00 pm–11:00 pm
COLOMBIA		
Bogotá	6.030	8:00 pm–Midnight
Bogotá	6.183	8:00 pm–Midnight
Bogotá	11.825	8:00 pm–Midnight
COMORO ISLANDS		
Moroni	7.260	10:30 pm–Midnight
COOK ISLANDS		
Rarotonga	5.045	11:30 pm–Midnight
COSTA RICA		
San José	6.075	8:00 pm–Midnight

Location	Freq (MHz)	Actual Time
CUBA		
Havana	9.760	8:00 pm–10:00 pm
Havana	11.760	10:30 pm–Midnight
Havana	11.840	8:00 pm–10:00 pm
CZECHOSLOVAKIA		
Prague	5.930	8:00 pm–9:00 pm
		10:00 pm–11:00 pm
Prague	7.345	8:00 pm–9:00 pm
		10:00 pm–11:00 pm
Prague	9.540	8:00 pm–9:00 pm
		10:00 pm–11:00 pm
Prague	9.630	8:00 pm–9:00 pm
		10:00 pm–11:00 pm
Prague	11.990	8:00 pm–9:00 pm
		10:00 pm–11:00 pm
DOMINICAN REPUBLIC		
Santo Domingo	6.090	8:00 pm–11:00 pm
Santo Domingo	9.505	8:00 pm–11:00 pm
ECUADOR		
Quito	9.605	8:10 pm–Midnight
Quito	11.745	8:00 pm–Midnight
Quito	15.115	8:00 pm–Midnight
EL SALVADOR		
San Salvador	5.980	8:00 pm–Midnight
San Salvador	9.555	8:00 pm–Midnight
ETHIOPIA		
Addis Ababa	9.725	11:00 pm–11:25 pm
FIJI ISLANDS		
Suva	3.230	10:45 pm–Midnight
Suva	6.005	8:00 pm–8:45 pm
FINLAND		
Helsinki	9.585	9:00 pm–10:00 pm

Location	Freq (MHz)	Actual Time

GABON

Libreville	3.300	11:30 pm–Midnight
Libreville	4.777	11:30 pm–Midnight

GERMANY (Democratic Rep.)

Berlin	5.955	8:00 pm–8:45 pm
		9:30 pm–10:15 pm
		10:30 pm–11:15 pm
Berlin	6.080	10:30 pm–11:15 pm
Berlin	6.165	10:30 pm–11:15 pm
Berlin	9.730	8:00 pm–8:45 pm
		9:30 pm–10:15 pm

GERMANY (Federal Rep.)

Cologne	6.075	10:45 pm–11:00 pm
		11:30 pm–Midnight
Cologne	6.145	11:30 pm–Midnight
Cologne	7.225	11:30 pm–Midnight
Cologne	7.235	10:45 pm–11:00 pm
Cologne	9.545	11:30 pm–Midnight
Cologne	9.565	11:30 pm–Midnight
Cologne	9.735	8:30 pm–10:00 pm

GREAT BRITAIN

London	6.110	8:00 pm–Midnight
London	7.130	8:00 pm–Midnight
London	9.410	11:00 pm–Midnight
London	9.510	8:00 pm–10:30 pm

GUATEMALA

Guatemala City	6.180	8:00 pm–11:00 pm
Guatemala City	9.505	8:00 pm–11:00 pm

GUYANA

Georgetown	3.265	8:00 pm–9:45 pm

HAITI

Cap Haitien	9.770	8:00 pm–10:30 pm
Cap Haitien	11.835	8:00 pm–10:30 pm

HOLLAND

Hilversum	11.730	9:00 pm–10:20 pm

Location	Freq (MHz)	Actual Time
HONDURAS (British)		
Belize	3.300	8:00 pm–Midnight
HONDURAS (Republic)		
Tegucigalpa	4.820	8:00 pm–Midnight
Tegucigalpa	6.050	8:00 pm–Midnight
HUNGARY		
Budapest	9.833	8:00 pm–8:30 pm
		10:00 pm–10:30 pm
Budapest	11.910	8:00 pm–8:30 pm
		10:00 pm–10:30 pm
Budapest	15.165	8:00 pm–8:30 pm
		10:00 pm–10:30 pm
Budapest	17.840	8:00 pm–8:30 pm
Budapest	21.685	8:00 pm–8:30 pm
INDIA		
New Delhi	15.235	8:00 pm–8:15 pm
ITALY		
Rome	9.575	8:00 pm–8:30 pm
Rome	11.810	8:00 pm–8:30 pm
Rome	15.330	10:45 pm–11:15 pm
Rome	17.795	10:45 pm–11:15 pm
JAPAN		
Tokyo	17.785	8:00 pm–8:15 pm
		9:00 pm–9:15 pm
		10:00 pm–10:15 pm
		11:00 pm–11:15 pm
Tokyo	17.855	8:00 pm–8:15 pm
		9:00 pm–9:15 pm
		10:00 pm–10:15 pm
		11:00 pm–11:15 pm
Tokyo	17.880	8:00 pm–8:15 pm
		9:00 pm–9:15 pm
		10:00 pm–10:15 pm
		11:00 pm–11:15 pm
KENYA		
Nairobi	4.915	10:00 pm–Midnight

Location	Freq (MHz)	Actual Time
KOREA (Dem. People's Rep.)		
Pyongyang	6.540	9:00 pm—10:00 pm
Pyongyang	15.150	9:00 pm—10:00 pm
KOREA (Republic of)		
Seoul	15.430	10:00 pm—10:30 pm
KUWAIT		
Kuwait	17.750	11:00 pm—Midnight
LESOTHO		
Maseru	4.800	11:00 pm—Midnight
LUXEMBOURG		
Villa Louvigny	6.090	8:00 pm—9:00 pm
MALAWI		
Zomba	3.380	11:00 pm—Midnight
Zomba	5.995	11:00 pm—Midnight
MALAYSIA		
Kuala Lumpur	9.660	8:00 pm—8:30 pm
Kuala Lumpur	11.900	8:00 pm—8:30 pm
Kuala Lumpur	15.280	8:00 pm—8:30 pm
MAURITIUS		
Forest Side	9.710	9:00 pm—Midnight
MEXICO		
Mexico City	11.770	8:00 pm—Midnight
Mexico City	17.835	8:00 pm—9:00 pm
Mexico City	21.705	8:00 pm—10:00 pm
MOZAMBIQUE		
Lourenco Marques	6.050	11:00 pm—Midnight
NEPAL		
Katmandu	9.595	8:20 pm—11:00 pm
NETHERLANDS ANTILLES		
Bonaire	11.820	8:00 pm—8:30 pm

Location	Freq (MHz)	Actual Time
NEW CALEDONIA		
Noumea	3.355	8:00 pm–9:00 pm
Noumea	7.170	8:00 pm–9:00 pm
NEW GUINEA		
Port Moresby	9.520	8:00 pm–Midnight
Port Moresby	11.880	8:00 pm–Midnight
NEW ZEALAND		
Wellington	15.110	8:00 pm–Midnight
Wellington	17.770	8:00 pm–Midnight
NICARAGUA		
Managua	11.875	8:00 pm–Midnight
NORWAY		
Oslo	9.550	8:00 pm–8:30 pm
Oslo	9.645	8:00 pm–8:30 pm
PAKISTAN		
Karachi	21.590	11:45 pm–Midnight
PANAMA		
David	6.045	8:00 pm–Midnight
PARAGUAY		
Asunción	5.975	8:00 pm–10:00 pm
Asunción	15.210	8:00 pm–10:00 pm
PHILIPPINES		
Manila	15.300	8:30 pm–9:00 pm
Manila	15.440	8:00 pm–Midnight
Manila	17.810	9:00 pm–Midnight
Manila	21.515	8:00 pm–Midnight
POLAND		
Warsaw	6.035	9:00 pm–10:30 pm
Warsaw	6.135	9:00 pm–10:30 pm
Warsaw	7.285	9:00 pm–10:30 pm
Warsaw	11.815	9:00 pm–10:30 pm
Warsaw	11.840	9:00 pm–10:30 pm

Location	Freq (MHz)	Actual Time
POLAND (cont)		
Warsaw	15.120	9:00 pm—10:30 pm
Warsaw	15.275	9:00 pm—10:30 pm
PORTUGAL		
Lisbon	6.025	10:45 pm—11:30 pm
Lisbon	11.935	9:00 pm—9:45 pm
		10:45 pm—11:30 pm
REUNION		
St. Denis	4.807	9:30 pm—Midnight
RHODESIA		
Salisbury	3.396	11:00 pm—Midnight
ROMANIA		
Bucharest	5.980	8:30 pm—9:30 pm
		10:00 pm—Midnight
Bucharest	6.150	8:30 pm—9:30 pm
		10:00 pm—10:30 pm
		11:30 pm—Midnight
Bucharest	6.190	8:30 pm—9:30 pm
		10:00 pm—10:30 pm
		11:30 pm—Midnight
Bucharest	9.510	8:30 pm—9:30 pm
		10:00 pm—10:30 pm
		11:30 pm—Midnight
Bucharest	9.570	8:30 pm—9:30 pm
		10:00 pm—10:30 pm
		11:30 pm—Midnight
Bucharest	9.690	8:30 pm—9:30 pm
		10:00 pm—10:30 pm
		11:30 pm—Midnight
Bucharest	11.940	8:30 pm—9:30 pm
		10:00 pm—10:30 pm
		11:30 pm—Midnight
RWANDA		
Kigali	6.055	10:30 pm—Midnight
SAUDI ARABIA		
Mecca	11.855	11:30 pm—Midnight

Location	Freq (MHz)	Actual Time
SEYCHELLES		
Victoria	11.920	9:45 pm–10:15 pm
Victoria	15.185	9:45 pm–10:15 pm
SINGAPORE		
Singapore	5.052	8:00 pm–Midnight
Singapore	11.940	8:00 pm–Midnight
SOLOMON ISLANDS		
Honiara	9.775	8:00 pm–10:00 pm
SOUTH AFRICA		
Johannesburg	6.080	8:00 pm–10:30 pm
Johannesburg	7.270	11:15 pm–Midnight
Johannesburg	9.525	11:15 pm–Midnight
Johannesburg	9.560	8:00 pm–10:30 pm
Johannesburg	9.695	8:00 pm–10:30 pm
Johannesburg	11.970	8:00 pm–10:30 pm
Johannesburg	15.220	11:15 pm–Midnight
Johannesburg	17.805	11:15 pm–Midnight
SPAIN		
Madrid	6.140	8:00 pm–11:00 pm
SRI LANKA (Ceylon)		
Colombo	6.075	8:00 pm–9:00 pm
Colombo	9.720	8:00 pm–11:30 pm
Colombo	15.120	8:00 pm–9:30 pm
SWEDEN		
Stockholm	6.175	9:00 pm–9:30 pm
Stockholm	11.705	10:30 pm–11:00 pm
SWITZERLAND		
Berne	6.120	8:30 pm–9:00 pm
		11:00 pm–11:30 pm
Berne	9.535	8:30 pm–9:00 pm
		11:00 pm–11:30 pm
Berne	9.750	8:30 pm–9:00 pm
Berne	11.715	8:30 pm–9:00 pm

Location	Freq (MHz)	Actual Time
TANZANIA		
Dar es Salaam	4.785	10:45 pm—11:45 pm
Dar es Salaam	6.105	10:45 pm—11:45 pm
THAILAND		
Bangkok	7.115	11:15 pm—Midnight
Bangkok	11.910	11:15 pm—Midnight
UNITED ARAB REPUBLIC (Egypt)		
Cairo	9.475	9:00 pm—10:00 pm
UNITED NATIONS		
New York, N.Y.	11.155	9:30 pm—9:45 pm
New York, N.Y.	17.830	9:30 pm—9:45 pm
UNITED STATES OF AMERICA		
Belmont, Calif.	9.670	11:30 pm—Midnight
Belmont, Calif.	15.280	11:30 pm—Midnight
Washington, D.C. (VOA)	3.980	10:00 pm—Midnight
Washington, D.C. (VOA)	5.955	10:00 pm—Midnight
Washington, D.C. (VOA)	5.965	10:00 pm—Midnight
Washington, D.C. (VOA)	5.995	10:00 pm—Midnight
Washington, D.C. (VOA)	6.015	11:00 pm—Midnight
Washington, D.C. (VOA)	6.160	10:00 pm—Midnight
Washington, D.C. (VOA)	7.195	11:00 pm—Midnight
Washington, D.C. (VOA)	7.200	10:00 pm—Midnight
Washington, D.C. (VOA)	7.280	11:00 pm—Midnight
Washington, D.C. (VOA)	9.530	11:00 pm—Midnight
Washington, D.C. (VOA)	9.635	10:00 pm—Midnight
Washington, D.C. (VOA)	9.740	10:00 pm—Midnight
Washington, D.C. (VOA)	9.750	11:00 pm—Midnight
Washington, D.C. (VOA)	11.760	10:00 pm—Midnight
U.S.S.R.		
Moscow	9.660	8:00 pm—10:30 pm 11:00 pm—Midnight
Moscow	9.668	8:00 pm—10:30 pm 11:00 pm—Midnight
Moscow	11.850	8:00 pm—10:30 pm 11:00 pm—Midnight
Moscow	11.870	8:00 pm—10:30 pm 11:00 pm—Midnight

Location	Freq (MHz)	Actual Time
		$\mathcal{E} \, 5 \, 7$
U.S.S.R. (cont)		
Moscow	11.900	8:00 pm–10:30 pm
		11:00 pm–Midnight
Moscow	11.920	8:00 pm–10:30 pm
		11:00 pm–Midnight
Moscow	15.210	8:00 pm–10:30 pm
		11:00 pm–Midnight
Moscow	17.730	8:00 pm–10:30 pm
		11:00 pm–Midnight
WINDWARD ISLANDS		
Grenada	11.975	8:00 pm–9:15 pm
ZAIRE		
Kinshasa	7.115	8:00 pm–10:00 pm
Kinshasa	15.265	8:00 pm–10:00 pm
ZAMBIA		
Lusaka	6.165	9:55 pm–Midnight
Lusaka	7.240	9:55 pm–Midnight

Clandestine Shortwave Broadcasting Stations

Location	Identification	Freq (MHz)	Transmission Period (EST)
Burma	"Voice of the People of Burma" (Myama-Pye Pey-Thu Ah-Than)		Burmese Language:
		5.114	*7:30 pm–8:30 pm
		5.114	*7:00 am–8:00 am
			Mandarin Language:
		5.114	†7:30 pm–8:30 pm
		5.114	†7:00 am–8:00 am

The communist clandestine radio "Voice of the People of Burma" was first heard in April, 1971. A Burmese Communist Party Central Committee statement announcing the radio's inauguration established its avowedly pro-Peking theme and stated that it will pursue a line of revolutionary armed struggle against the present government with its ultimate aim being the "realization of the communist system in Burma." Although the station has not announced its location, technical observations indicate that it is situated in China's southwest Yunnan Province.

---

*Tues., Thurs., Fri., Sun.
†Sat.

Location	Identification	Freq (MHz)	Transmission Period (EST)
Cambodia	"Voice of the		Cambodian Language:
	National United Front of Kampuchea" (Vitayu	4.988, 7.010, 9.988, 10.080, 12.006	6:00 pm–7:00 pm
	Phsay Samleng Ronnacse Ruab Ruam Cheat	4.988, 9.988, 10.080, 12.006	7:00 pm–8:00 pm
	Kampuchea)	4.988, 9.988, 10.080, 12.006	11:00 pm–Midnight
		4.988, 7.010, 9.988, 10.080, 12.006	6:30 am–7:30 am
		4.988, 9.988, 10,080, 12.006	9:30 am–12:30 pm

Broadcasts of this pro-Sihanouk radio were officially inaugurated in August, 1970. Established by decision of the "Ministry of Information and Propaganda of the RGNUC" (Sihanouk's Peking-based government-in-exile), the radio stresses "Cambodia liberation" and the legitimacy of the RGNUC, and vehemently opposes the U.S. and Lon Nol-Sirik Matak "cliques." Although claiming operation from within liberated areas of Cambodia, technical observations indicate that all broadcasts emanate from locations in North Vietnam.

Location	Identification	Freq (MHz)	Transmission Period (EST)
Cyprus	"Voice of Truth" (Radiofonikos Stathmos I Foni Tis Alithias)		Greek Language (To Cyprus and Greece):
		7.335	11:30 pm–Midnight
			2:00 am–2:30 am
			4:45 am–5:15 am
			11:00 am–11:30 am
			1:00 pm–1:40 pm
			2:00 pm–2:45 pm
			3:00 pm–4:00 pm
		9.775	5:20 am–5:50 am
			11:35 am–12:05 pm

The communist transmissions "Voice of Truth" and "Our Radio" were first heard in March, 1958. They oppose the present regime in Cyprus, Greece, and Turkey, calling for an end to United States influence in those countries. Technical observations indicate the transmissions are radiated from East Germany and Romania.

Location	Identification	Freq (MHz)	Transmission Period (EST)
Greece	"Voice of Truth" (Radio-fonikos Stathmos I Foni Tis Alithias) (See Cyprus for details.)		
Iran	"National Voice of Iran" (Seday-e Melli Iran)	6.025	Azerbaijani Language: 12:45 pm–1:00 pm
			Kurdish Language: 1:00 pm–1:15 pm
			Persian Language: 12:30 pm–12:45 pm

The only communist clandestine station still broadcasting from Soviet territory, the "National Voice of Iran" strongly condemns United States influence in Iran and has called for the violent overthrow of the Shah. Technical observations indicate that this broadcaster employs the same transmitters that are used to carry Baku's international and domestic services. The station was first heard in April, 1959.

Location	Identification	Freq (MHz)	Transmission Period (EST)
Iran	"Radio Iran Courier" (Radio Peyk-e Iran)	9.560, 11.410, 11.695	Azerbaijani Language: 10:00 am–10:30 am
		11.410, 11.695	11:50 am–12:20 pm
		9.560, 11.410, 11.695	Kurdish Language: 9:30 am–10:00 am
		11.410, 11.695	11:20 am–11:50 am
		9.560, 11.410, 11.695	Persian Language: 10:30 am–11:20 am
		11.410, 11.695	12:20 pm–1:10 pm

This radio voice assails the Iranian government for its domestic, regional, and foreign policies, including its close association with the United States. Although no location is announced, broadcasts of "Radio Iran Courier" are believed to have emanated from Bulgaria since 1963.

Location	Identification	Freq (MHz)	Transmission Period (EST)
Korea (North)	"Voice of the Revolutionary Party for Reunification" (Tongil Hyongmyong-Tang Moksori)	4.558	Korean Language: 7:00 am–9:00 am

This communist-controlled station was first heard in August, 1970. The transmissions adhere to Pyongyang's revolutionary propaganda line, calling on the South Korean people to join the struggle against the "U. S. Imperialists," and propagandizes the political program of the "South Korean Revolutionary Party for Reunification." Technical observations indicate that its transmissions emanate from North Korea —probably at the facilities of Pyongyang's Korean Central Broadcasting Station.

Location	Identification	Freq (MHz)	Transmission Period (EST)
			French Language:
Laos	"Radio Pathet Lao" (Thi Ni Withayu Kachai Siang Fai Pathet Lao)	4.660, 6.200, 7.310	8:00 pm–8:15 pm
		6.200, 7.310, 8.635, 8.660	12:45 am–1:00 am
		4.660, 6.200, 7.310	Midnight–12:15 am
			Laotian Language:
		4.660, 6.200, 7.310	5.30 pm–7:00 pm
		4.660, 6.200, 7.310	8:30 pm–9:30 pm
		6.200, 7.310, 8.635, 8.660	10:45 pm–12:45 am
		4.660, 6.200, 7.310, 7.480	4:00 am–6:00 am
		4.660, 6.200, 7.310	8:00 am–11:00 am
			Vietnamese Language:
		4.660, 6.200, 7.310	8:15 pm–8:30 pm
		4.660, 6.200, 7.310, 7.480	6:45 am–7:00 am

Location	Identification	Freq (MHz)	Transmission Period (EST)

This communist-controlled radio station is the mouthpiece of the "Neo Lao Hak Sat" (NLHS), the leftist faction in Laos. Its broadcasts were first heard in August, 1960. "Radio Pathet Lao" severely criticizes Souvanna Phouma's government as being a U.S. "puppet," and generally follows Hanoi's propaganda line on developments in Indochina. Technical observations indicate that the station is using facilities located in North Vietnam.

			Laotian Language:
Laos	"Radio of the Patriotic	6.273, 8.600	6:00 am–7:00 am
	Neutralist Forces" (Thi	6.273, 8.600	8:00 am–9:00 am
	Ni Satani Withayu	6.273, 8.600	6:15 pm–7:15 pm
	Kachai Siang Heng		
	Pathet Lao)	6.273, 8.600	Midnight–1:00 am

Although ostensibly representing the neutralist factions in Laos, this communist-controlled station echoes the propaganda line taken by "Radio Pathet Lao," mouthpiece of the leftist NLHS. The station was first heard in December, 1960, and originally identified itself as "Radio of the Laotian Kingdom." Although claiming to operate from Khang Khay, Laos, technical observations indicate that it utilizes facilities located in North Vietnam.

			Malay Language
Malaysia	"Voice of the	7.305, 11.828	5:00 pm–5:40 pm
	Malayan Revolu-	7.305, 11.828	11:30 pm–12:10 am
	tion" (Ini-Lah	7.305, 11.828	6:00 am–6:40 am
	Suara Revolusi		
	Malaya)		Mandarin Language:
		7.305, 11.828	5:45 pm–6:15 pm
		7.305, 11.828	12:15 am–12:45 am
		7.305, 11.828	6:45 am–7:15 am

This communist-controlled radio began broadcasting in November, 1969, in the Malay and Mandarin languages. The station follows a Peking-oriented line and vehemently attacks the present Malaysian and Singaporean governments, specifically the Razak and Lee Kuan Yew "cliques." Although claiming to be "the Malayan people's own

Location	Identification	Freq (MHz)	Transmission Period (EST)

radio station," and according to Peking, a "people's radio set up by the Communist Party of Malaya," technical observations indicate that the station is located in China's Hunan Province.

Location	Identification	Freq (MHz)	Transmission Period (EST)
			Portuguese Language:
Portugal	"Radio Free	15.483	2:00 am–2:30 am
	Portugal"	12.005, 14.440,	7:00 am–7:30 am
	(Radio	14.955, 15.483	
	Portugal	11.505	1:00 pm–3:00 pm
	Livre)		
		8.332, 9.455,	6:20 pm–6:50 pm
		11.505	

"Radio Free Portugal" and "Radio Independent Spain" are communist-controlled broadcasters who emphasize opposition to the Caetano and Franco regimes. "Radio Free Portugal," first heard in 1962, does not announce its location. However, "Radio Independent Spain," the oldest active communist clandestine station, began broadcasting to Spain in 1941, and claims to operate from the Pyrenees. Technical observations indicate that the transmitters are located in Romania.

Location	Identification	Freq (MHz)	Transmission Period (EST)
			Spanish Language:
Spain	"Radio Independent	7.690, 10.110,	1:00 am–1:55 am
	Spain"	12.140, 14.485	
	(Radio Espana Inde-	15.365	7:35 am–7:55 am
	pendiente)	12.140, 14.485,	8:00 am–9:00 am
		15.505, 17.660	
	(See Portugal for	7.690, 10.110,	11:00 am–1:00 pm
	details.)	12.140, 14.485	
		7.690, 9.430,	1:00 pm–6:10 pm
		10.110, 12.140	
		9.833	5:30 pm–5:50 pm

Location	Identification	Freq (MHz)	Transmission Period (EST)
			Laotian Language:
Thailand	"Voice of the People	6.033, 9.423	7:00 am–7:40 am
	of Thailand" (Thi Ni	6.033, 9.423	9:30 am–10:10 am
	Sathani Vithayu	6.033, 9.423	11:00 pm–11:40 pm
	Sieng Prachachon		
	Heang Prathet Thai)		Thai Language:
		6.033, 9.423	5:00 am–5:40 am
		6.033, 9.423	8:00 am–8:40 am
		6.033, 9.423	10:30 am–11:10 am
		6.033, 9.423	6:00 pm–6:40 pm
		6.033, 9.423	10:00 pm–10:40 pm
		6.033, 9.423	Midnight–12:40 am

This broadcaster is critical of the Thai and U.S. Governments, and propagandizes the position of the outlawed Communist Party of Thailand as representative of the true interests of the Thai people. First heard in April, 1962, this clandestine radio initially was reported operating from communist-held territory in central Laos; however, technical observations indicate that its transmitters are located in China's Yunnan Province.

Location	Identification	Freq (MHz)	Transmission Period (EST)
			Turkish Language:
Turkey	"Our Radio"	9.500	1:00 am–1:30 am
	(Bizim Radyo)	9.730	2:00 am–2:30 am
	(See Cyprus for details.)	9.730	5:15 am–5:45 am
		9.500	9:45 am–10:15 am
		9.730	1:00 pm–1:30 pm
		5.915	3:00 pm–4:00 pm

Location	Identification	Freq (MHz)	Transmission Period (EST)
			English Language:
Vietnam	"Liberation	7.470, 10.010	5:30 am–5:45 am
(North)	Radio"	7.470, 10.010	9:30 am–10:00 am
	(Dai Phat	7.470, 10.010	5:00 pm–5:30 pm
	Thanh Giai	7.470, 10.010	6:30 pm–7:00 pm
	Phong)		
			French Language:
		7.470, 10.010	9:00 am–9:30 am
		12.110, 14.993	3.00 pm–3:30 pm
		7.470, 10.010	6:00 pm–6:30 pm
			Vietnamese Language:
		7.420, 10.225	5:00 pm–9:30 pm
		7.420, 10.225	10:30 pm–1:00 am
		10.110, 12.110, 14.990	2:00 pm–4:00 pm

This communist radio began broadcasting in February, 1962, identifying itself as "Liberation Radio, the Voice of the National Front for the Liberation of South Vietnam." On June 26, 1969, its identification was changed to "Liberation Radio, The Voice of the Provisional Revolutionary Government of the Republic of South Vietnam." The station's anti-U. S. broadcasts call for the overthrow of the present Saigon government. Although purportedly operating in a communist-controlled area of South Vietnam, technical observations indicate that it is using the transmitting facilities of Hanoi's official radio, "Voice of Vietnam."

STATION LOG

LOCATION	FREQ	TIME	REMARKS
Russia	12 Mhz / 88+62	12:45 MST / Sun-Fri 25	Find out it!
VOA West	9.76? / D-70+70	1:15 PM MST / for 2	Fed!
VOA - West	6.040	2-4 PM MST / Scan	Dates
Radio Netherlands Hilversum, Holland	6+ M hg	5:50 M.	L.K.Saul - prog. AFR / Mutual
	11.8	1-2 pm	
BBC	6.0	5:00AM	Richard Baker (moderator)
Russian (?) language new	21 Mh	1304-1-2-80 / 5 05428	
BBC - English Relay / Canadian Relay	10.3 mhz / D-7421 / 6Khz / D 7347 66	1-2-80 / 1725 MST / Mon 1/7/80	BBC
Colonge W. Germany	86+24	1850 MST / Mon 1/7/80	Voice of Germany 530-550 / 5 9631 6100
VOA	11.9 / D 86+24	1850 MST	2/1 @ 1900 11,705
BBC		1900 MST	News / 6100 6175

STATION LOG

LOCATION	FREQ	TIME	REMARKS
S. Africa?	E 4512?	1/7/80	
	D 73+26 17.70 ? MHz	1950	
Netherlands (news + commentary)	E 10 ± MHz	1/31/80	
Rome Service - news + comment.	11 MHz / D 73+39	9/2	
Radio Moscow	R 18 MHz	1-31-80	
	E 40 + 31.5	1976	
Radio Japan Tokyo	E 40 + 33	2-3-80 1915	news, commentary
Radio Canada	E 40 + 37	2-23-80 1301	news
Russian Language	E 60+5 6½	4-27-80 135-9 MDT	Russian language newscast
BBC (?) News	12.095~(?)	4-25-82 8/14/81 ST	12 mtr 12.0 MHz 7 64 Lg BBC news
VOA	M 970 (?)	4.25.82 1032 MDT	VOA - 8 something Lg pg

STATION LOG											
LOCATION	FREQ	TIME	REMARKS								

STATION LOG

LOCATION	FREQ	TIME	REMARKS

STATION LOG

LOCATION	FREQ	TIME	REMARKS

STATION LOG

LOCATION	FREQ	TIME	REMARKS